Keio University International Symposia
for Life Sciences and Medicine 11

Springer

Tokyo
Berlin
Heidelberg
New York
Hong Kong
London
Milan
Paris

A. Kaneko (Ed.)

The Neural Basis of Early Vision

With 65 Figures, Including 10 in Color

 Springer

Akimichi Kaneko, M.D., Ph.D.
Professor
Department of Physiology
School of Medicine, Keio University
35 Shinanomachi, Shinjuku-ku, Tokyo 160-8582, Japan

ISBN 4-431-00459-9 Springer-Verlag Tokyo Berlin Heidelberg New York

Library of Congress Cataloging-in-Publication Data applied for.

Printed on acid-free paper

Typesetting: SNP Best-set Typesetter Ltd., Hong Kong
Printing and binding: Nikkei Printing Inc., Japan
SPIN: 10907689

Foreword

This volume of the *Keio University International Symposia for Life Sciences and Medicine* contains the proceedings of the twelfth symposium held under the sponsorship of the Keio University Medical Science Fund. The fund was established by the generous donation of Dr. Mitsunada Sakaguchi. The Keio University International Symposia for Life Sciences and Medicine constitute one of the core activities sponsored by the fund, of which the objective is to contribute to the international community by developing human resources, promoting scientific knowledge, and encouraging mutual exchange. Each year, the Committee of the International Symposia for Life Sciences and Medicine selects the most significant symposium topics from applications received from the Keio medical community. The publication of the proceedings is intended to publicize and distribute the information arising from the lively discussions of the most exciting and current issues presented during the symposium. On behalf of the Committee, I am most grateful to Dr. Mitsunada Sakaguchi, who made the series of symposia possible. We are also grateful to the prominent speakers for their contribution to this volume. In addition, we would like to acknowledge the efficient organizational work performed by the members of the program committee and the staff of the Medical Science Fund.

Naoki Aikawa, M.D., D.M.Sc., F.A.C.S.
Chairman
Committee of the International Symposia
for Life Sciences and Medicine

V

The 12th Keio University International Symposium for Life Sciences and Medicine
The Neural Basis of Early Vision   *September 2–4, 2002*

Preface

The twelfth Symposium for Life Sciences and Medicine of the Keio University Medical Science Fund was held under the title of "The Neural Basis of Early Vision." In the symposium we discussed recent advances in the visual neurosciences of the vertebrate retina and the visual cortex.

Retina research at Keio University began in the mid-1950s, when the late Professor Tsuneo Tomita started to analyze the origin of the electroretinogram. In the mid-1960s Tomita and his group succeeded in making an intracellular recording from single photoreceptors and showed that light hyperpolarizes photoreceptors. Since their discovery, studies on phototransduction and signal processing in the retina have flourished worldwide, and a vast amount of knowledge has accumulated in the last 40 years. However, old questions on signal processing in the retina have been reexamined recently, and a lively discussion on these topics took place in this symposium.

Modern studies on information processing in the visual cortex were initiated by Professors Hubel and Wiesel in about 1960. They also demonstrated that the neural circuit of the visual cortex is highly flexible particularly during the critical period of development in young animals. Much research followed theirs and has elucidated the function of the visual cortex in recent decades. Participants at the Symposium were encouraged by the presence of Professor Torsten N. Wiesel, who shared the 1981 Nobel Prize in Physiology or Medicine with Professor David H. Hubel for their discoveries concerning information processing in the visual system.

A large number of researchers in Japan have met almost every year since 1982 to exchange information on their studies of the visual system, and in 1997 they established Vision Forum. This year, Vision Forum joined us in this symposium. I thank Professor Masao Tachibana, the president of Vision Forum, and Professor Shu-Ichi Watanabe, the organizer of the 2002 Vision Forum, for contributing to the twelfth Symposium for Life Sciences and Medicine by organizing a joint meeting with Vision Forum. I also want to mention the support of Professor Shiro Usui, principal investigator of the project "Neuroinformatics in Vision (NRV)," which is being carried out as Target-Oriented R&D for Brain Science at the MEXT.

The Symposium was successful because of the large body of participants—a total of 236, including 18 from abroad (14 from the United States, 2 from France, and 1 each from Germany and Finland). Thirty-one papers were presented orally and 56 poster presentations were made. I strongly believe that this symposium has contributed significantly to the advancement of visual neuroscience in Japan and in the world.

The symposium would not have been possible without the support of the Keio University Medical Science Fund established by the generous donation of Dr. and Mrs. Mitsunada Sakaguchi. I wish to express my deep appreciation to the founder, and I am grateful for all the efforts of the staff members of the Fund. Finally, I thank Springer-Verlag, Tokyo, for their excellent work in publishing this volume.

Akimichi Kaneko, M.D., Ph.D.
Editor, Twelfth Keio University International Symposium
for Life Sciences and Medicine
on the Neural Basis of Early Vision

Contents

List of Contributors

Psychophysics to Biophysics: How a Perception Depends on Circuits, Synapses, and Vesicles

Narender Dhingra, Robert Smith, and Peter Sterling

Key words. Ganglion cell, Psychophysics, Biophysics

One stimulus that we detect very efficiently is a small square that just covers the dendritic field of a brisk-transient ganglion cell. Since this cell type is the most sensitive of the geniculo-striate projecting ganglion cells, it may largely mediate this behavior. To compare the neuron's sensitivity to that of psychophysical detection, we measured its visual threshold with a method borrowed from psychophysics. Recordings were extracellular from a mammalian brisk-transient ganglion cell (guinea pig), whose impulse response was nearly identical to that of the primate brisk-transient cell. The stimulus, a 100-ms spot covering the receptive field center, was detected by an "ideal observer" with knowledge of the spike patterns from 100 trials at each contrast. Based on this knowledge, the ideal observer used a single-interval, forced-choice procedure to predict the stimulus contrast on 100 additional trials. Brisk-transient cells at 37°C detected contrasts as low as 0.8% (mean $\pm$ SEM = 2.8% $\pm$ 0.2) and discriminated between contrast increments with about 40% greater sensitivity. These thresholds are the same as human psychophysical thresholds for comparable stimuli, suggesting that across many levels of noisy central synapses, little or no information is lost. Recording intracellularly, we found the detection threshold of the ganglion cell's graded potential to be about half that of the spike response, implying a considerable loss in converting the signals from analog to digital. To reach detection threshold, the ganglion cell needed ~1000 quanta, and to respond at full contrast, it needed ~2000 quanta. Since the ribbon synapses that contact this cell contain an aggregate of ~10^5 releasable vesicles, the safety factor for this circuit seems to be about 5.

Department of Neuroscience, University of Pennsylvania, Philadelphia, PA 19104, USA

Phototransduction in Rods and Cones, and Beyond

King-Wai Yau

Key words. Retina, Photoreceptor, Phototransduction

Retinal rods and cones hyperpolarize to light as a result of a light-triggered cyclic guanosine monophosphate (cGMP) signaling pathway. In this phototransduction process, light activates rhodopsin, which, via the G protein transducin, stimulates a phosphodiesterase to increase cGMP hydrolysis, causing the intracellular free cGMP level to fall and cGMP-gated, nonselective cation channels that are open in darkness to close, hence producing the hyperpolarization. This closure of the cGMP-gated channels, which are Ca^{2+} permeable, leads to a decrease in the free Ca^{2+} concentration in the outer segment. This Ca^{2+} decrease in the light activates multiple negative-feedback mechanisms to produce light adaptation.

Unlike rods and cones, the majority of invertebrate photoreceptors depolarize to light. Work on *Drosophila* and *Limulus* photoreceptors suggests that phototransduction in these cells involves a phospholipase C pathway, even though the second messenger controlling the ion channels in the final step still remains unclear. The striking difference in phototransduction mechanism between these cells and rods and cones raised the possibility of a fundamental dichotomy between vertebrate and invertebrate photoreceptors, or perhaps between hyperpolarizing and depolarizing photoreceptors. This notion, however, is untrue. Photoreceptors in the lizard parietal eye, which we are studying, depolarize to light under dark-adapted conditions, but they nonetheless use a cGMP pathway for phototransduction. In this case, light *inhibits* phosphodiesterase activity to lead to a cGMP *rise* and the *opening* of nonselective cation channels, hence producing the depolarization. There is also an unusual photoreceptor in scallop that hyperpolarizes to light, and it, too, uses a cGMP signaling pathway for phototransduction. In this case, light elevates cGMP to open a cGMP-activated *potassium* channel, hence producing the hyperpolarization [1]. So far, all photoreceptors—vertebrate or invertebrate, hyperpolarizing or depolarizing—known to use cGMP for transduction are *ciliary* photoreceptors, whereas those known

Department of Neuroscience and Howard Hughes Medical Institute, Room 907, Preclinical Teaching Building, Johns Hopkins University School of Medicine, 725 North Wolfe Street, Baltimore, MD 21205, USA

to use a phospholipase C pathway for transduction are *rhabdomeric* (microvillous) photoreceptors. This may be a general principle. Occasional photoreceptors, such as individual photosensitive neurons in the invertebrate central nervous system, appear to be neither ciliary nor rhabdomeric, in which case this principle does not apply.

Remarkably, in addition to rods and cones, a subset (ca. 1%) of retinal ganglion cells now appears to be also *intrinsically* photosensitive. These cells express melanopsin, an opsin-like protein, and they project to brain centers involved in non-image-forming visual functions (circadian photoentrainment, pupillary light reflex, etc.) such as the suprachiasmatic nucleus, the intergeniculate leaflet, and the olivary pretectal nucleus. Whether melanopsin is the light-sensing pigment and what photo-transduction mechanism is used in these cells remain to be studied.

Reference

1. Gomez M, Nasi E (1995) Activation of light-dependent K^+ channels in ciliary invertebrate photoreceptors involves cGMP but not the IP3/Ca^{2+} cascade. Neuron 15:607–618

Efficiency of Phototransduction Cascade in Cones

Shuji Tachibanaki[1], Sawae Tsushima[1], and Satoru Kawamura[2]

Key words. Phototransduction, Cones, Visual pigment phosphorylation

Cone photoreceptors are less light sensitive than rods, and the photoresponse of a cone is much briefer than that of a rod. In previous studies, the phototransduction cascade in cones has been shown to be essentially the same as that of rods: there are rod- and cone-version of components of the cascade. For this similarity, it has been speculated that the difference between a rod and a cone photoresponse is due to a quantitative difference in the reaction(s) of the cascade between rods and cones. However, the actual difference has not been measured yet. To understand the molecular mechanisms characterizing cone photoresponses, we compared the reactions in the phototransduction cascade between rods and cones.

Rods and cones were purified in quantities large enough to do biochemical studies. The cells were obtained from the carp (*Cyprmus carpio*) retina with a stepwise density gradient using Percoll. The purified rod fraction contained almost no other kinds of cells other than rods, and the purified cone fraction contained a mixture of red-, green-, and blue-sensitive cones in the ratio of $3:\approx1:\approx1$. We prepared membrane preparations from these purified cells to quantify the phototransduction reactions biochemically in rods and cones.

The results showed that both transducin activation by a bleached pigment and cyclic guanosine monophosphate phosphodiesterase activation by an activated transducin molecule are less effective in the cone membranes. The lower amplification in cones thus accounts for lower light sensitivity in cones. Furthermore, we measured the time courses of visual pigment phosphorylation, which is the mechanism of quench of light-activated visual pigment. The result showed that the phosphorylation is much faster (>20) in the cone membranes. Since the quench was much faster in cones, this result explains the lower light sensitivity as well as the briefer photoresponse in cones.

[1] Department of Biology, Graduate School of Science, and [2] Graduate School of Frontier Biosciences, Osaka University, 1-1 Machikaneyama, Toyonaka, Osaka 560-0043, Japan

Voltage-Gated Ion Channels in Human Photoreceptors: Na$^+$ and Hyperpolarization-Activated Cation Channels

EI-ICHI MIYACHI and FUSAO KAWAI

Key words. Retina, Photoreceptor, Voltage-gated channel, Sodium channel, h channel

Introduction

A light stimulus hyperpolarizes photoreceptors in biochemical processes in the outer segment and reduces the release of neurotransmitter by decreasing a Ca^{2+} influx at their synaptic terminals [1–6]. The photovoltage is shaped by voltage-gated channels in the inner segment [7–11]. Major voltage-gated currents measured in vertebrate photoreceptors are an L-type Ca^{2+} current, a delayed rectifier K$^+$ current, a fast transient K$^+$ current, and a hyperpolarization-activated cation current (*h* current) [7, 9, 10, 12]. Although mammalian photoreceptors are commonly thought to be nonspiking neurons [7, 9–11, 13], electrophysiological recordings with suction electrodes show that the termination of a light stimulus induces spike-like current responses in monkey photoreceptors [14]. This raises the possibility that primate photoreceptors may be able to generate action potentials. However, this hypothesis still remains uncertain, as there are few voltage recordings from primate photoreceptors [11, 15, 16]. Using the patch-clamp technique, we examined whether human rod photoreceptors can elicit action potentials, and also investigated the role of the voltage-gated currents for rod voltage responses.

Among those voltage-gated currents, *h* currents (I$_h$) are known to underlie the anomalous voltage rectification in photoresponses of cold-blooded vertebrate photoreceptors [7, 9, 17]. Recently, I$_h$-like currents were also found in monkey photoreceptors [10]. This raises the possibility that primate photoreceptors may express I$_h$ as well as cold-blooded vertebrate photoreceptors and I$_h$ may modulate voltage responses of primate photoreceptors. Using the patch-clamp recording technique, we also examined whether human rod photoreceptors express I$_h$, and investigated a role of I$_h$ for rod's voltage responses.

Department of Physiology, Fujita Health University School of Medicine, Toyoake, Aichi 470-1192, Japan

Materials and Methods

Slice Preparation and Dissociation Procedure

A small piece (diameter less than 1 mm) of peripheral retina was excised from adult patients during the surgical procedure to reattach the retina and during foveal translocation surgery (for details, see Kawai et al. [18]). All experiments were performed in compliance with the guidelines of the Physiological Society of Japan and the Society for Neuroscience, the Declaration of Helsinki, and also approved by the ethics committee of our institute. In the present experiment we used the peripheral retina (eccentricity approximately 20 mm), and selectively recorded from rods. In this region the spatial density of rods is much higher than that of cones [19, 20].

Slices from adult human retina were cut at 200 μm and viewed on an Olympus upright microscope with differential interference contrast optics (×40 water-immersion objective). Because the retinal detachment and the intentionally detached retina from foveal translocation surgery are in a condition where the sensory retina is physically isolated from the pigment epithelium, the human retinae obtained in the present experiments would be in a similar condition to isolated retinae of various species used widely for the patch-clamping recordings [21, 22]. A normal appearance of the retinal slice (not shown) and a functional recovery of patients after the retinal surgeries indicate that there are no severe pathological conditions in the retina used in the present experiments.

The dissociation procedure of human rods is similar to that of monkey photoreceptors by Yagi and Macleish [10]. A small piece of peripheral human retina was incubated for 5 min at 37°C in a solution containing 7 units/ml of papain (Sigma, St. Louis, MO, USA) with no added Ca^{2+} and Mg^{2+}. The tissue was then rinsed twice with Ames' medium and triturated. Isolated cells were plated on the concanavalin A-coated glass coverslip.

Electrophysiological Recordings

Membrane currents of human rods were recorded in the whole-cell configuration [23] using a patch-clamp amplifier (Axopatch 200B, Axon Instruments) linked to a computer. The voltage clamp procedures were controlled by the pCLAMP software (Axon Instruments). Data were low-pass filtered (4-pole Bessel type) with a cut-off frequency of 5 kHz and then digitized at 10 kHz by an analog-to-digital interface.

Retinal slices or isolated rods were perfused at 1 ml/min with Ames' medium (which is buffered with 22.6 mM bicarbonate), equilibrated with 95% O_2-5% CO_2, and maintained at 37°C. In slice preparations, $CoCl_2$ (1 mM), picrotoxin (100 μM), and strychnine (1 μM) were added to the bath to block synaptic inputs [21]. In both retinal slices and isolated cells for experiments on I_h, 1 μM tetrodotoxin was added to the bath to block Na^+ action potentials [18]. During voltage-clamp recordings of I_h in rods, the bath solution contained 20 mM TEA and 1 mM $CoCl_2$ to block delayed rectifier K^+ currents and voltage-gated Ca^{2+} currents, respectively. The recording pipette was filled with K^+ solution [in mM; KCl, 140; $CaCl_2$, 1; ethylene glycol tetraacetic acid (EGTA), 5; hydroxyethylpiperazine ethanesulfonic acid (HEPES), 10], K^+ solution with a low concentration of chloride (potassium gluconate, 125; KCl, 15; CaCl$_2$, 1; EGTA, 5; HEPES, 10), or Cs$^+$ solution (CsCl, 140; CaCl$_2$, 1; EGTA, 5; HEPES, 10). For studies on I$_h$, patch pipette solutions for voltage-clamp experiments contained 140 mM KCl, and those for current-clamp experiments contained 125 mM potassium gluconate. The solution was adjusted with KOH or CsOH to pH 7.4. Pipette resistance was about 9 MΩ. All test substances were applied through the bath (for details, see Kawai et al. [8, 24]).

Results

Na$^+$ Action Potentials in Human Rods

At first, we chose retinal slices rather than dissociated cells in order to minimize the damage to the photoreceptors. We recorded the membrane voltage and currents from the perikarya of the photoreceptors using the whole-cell patch clamp. To identify the types of cells in the retinal slice, we added Lucifer yellow to the pipette solution. The inner segment, perikarya, and axon of the photoreceptor were clearly stained by Lucifer yellow; however, the outer segment did not stain, probably due to a partial loss of outer segments caused by the retinal detachment. The rod-shaped inner segment suggests that the recorded cell is a rod photoreceptor. No dye coupling between rods was observed ($n = 0/23$). In the present study we selectively recorded from rods.

To test whether human rods can generate action potentials, we recorded the voltage response of a rod to depolarizing current injection under current-clamp conditions. In order to remove inactivation of the voltage-gated channels, the membrane potential was hyperpolarized from the resting level (-58 ± 4 mV; mean $\pm$ SEM, $n = 13$) by injection of steady negative current. This condition is equivalent to a steady light stimulus, as the steady light keeps a rod hyperpolarized [20, 25, 26]. Surprisingly, depolarizing current steps (+60 and +80 pA) induced a single action potential. This spike was blocked by a voltage-gated Na$^+$ channel blocker, 1 μM tetrodotoxin (TTX), but not the Ca^{2+} channel blocker, 1 mM Co^{2+}. A similar result was obtained in rods of the retina excised for the surgery of the foveal translocation ($n = 3$, not shown). When recording pipettes filled with a low concentration of chloride were used, we observed continued spiking in two out of three rods and a single action potential in one rod. These action potentials were also blocked by 1 μM TTX. This suggests that they were Na$^+$ spikes rather than Ca^{2+} spikes.

Because a light stimulus hyperpolarizes photoreceptors by reducing the cyclic guanosine monophosphate (cGMP)-gated inward current in the outer segment [20, 25, 26], we also examined the effects of hyperpolarizing current steps of various intensities on the voltage responses of a rod. To mimic a light stimulus in darkness, we depolarized the rod to the dark resting potential (approximately -40 mV [11]) by injection of a steady positive current. At the termination of hyperpolarizing current steps (-80 and -100 pA), prominent action potentials were observed ($n = 3$). This suggests that the termination of the light stimulus might elicit Na$^+$ spikes in human rods. As the photocurrent is known to decay gradually with time [20, 25, 26], we also investigated the effects of the "off" time course of the hyperpolarizing current on voltage responses. The cessation of the current injection at the rates of +3.2 and 1.07 pA/ms induced marked action potentials ($n = 3$). As the off-phase of the hyperpolarizing current was slowed down (+0.64, +0.46, and +0.36 pA/ms), the spike amplitude decreased gradually, probably due to inactivation of the voltage-gated currents in the inner segment and/or axon. The spikes were blocked by 1 µM TTX. This strongly suggests that Na^+ action potentials would be elicited in human rods when a light is turned off.

To understand the mechanism underlying spike generations in human rods, we recorded their membrane currents under voltage-clamp conditions. A steady inward current of 250 pA was observed at a holding potential (V_h) of −100 mV. Depolarizing voltage steps induced a transient inward current immediately following the fast capacitive current. The transient inward current began to be activated at −50 mV, and was maximal at −40 mV. A similar current was recorded in 14 out of 21 rods in the retinae obtained from patients with retinal detachment, and in 3 out of 3 rods in the retina obtained by surgery for foveal translocation. Mean peak amplitude was 441 ± 20 pA ($n = 14$) in the retinal detachment and 452 ± 43 pA ($n = 3$) in the foveal translocation. However, when a V_h was −50 mV, depolarizing voltage steps did not evoke the transient inward current, probably due to inactivation of the current. This current was reversibly blocked by 1 µM TTX; however, 1 mM Co^{2+} was ineffective. This implies that the transient inward current is carried through voltage-gated Na^+ channels and that action potentials in human rods are Na^+ spikes. After the termination of the command pulse, a slowly decaying outward tail current was observed.

To relate the voltage-gated Na^+ currents to the fast action potentials in human rods, we analyzed their kinetics in detail. The decay phase of the Na^+ currents induced by depolarization to −40 mV could be fitted by a single exponential function of a time constant of 1.6 ms. The decaying time constant of the Na^+ currents was maximal at −50 mV (2.3 ± 0.1 ms, $n = 4$), and decreased as the membrane voltage was depolarized. The kinetics of the voltage-gated Na^+ current is much faster than that of voltage-gated Ca^{2+} currents in other preparations [27–29], suggesting that the voltage-gated Na^+ currents are responsible for generating fast action potentials in human rods. The activation and inactivation curves of voltage-gated Na^+ currents were fitted by a single Boltzmann function. The half-activation voltage was 35 mV, and the half-inactivation voltage was −65 mV ($n = 4$). These values are similar to those of voltage-gated Na^+ currents in other preparations [27–29].

Hyperpolarization-Activated Cation Current (h Current) in Human Rods

We also explored the membrane currents activated by hyperpolarization from the resting potential of rods in darkness (−40 mV) [11]. At a V_h of −40 mV, hyperpolarizing voltage steps evoked a slow inward current. After the termination of the command pulse, an inward tail current was also recorded. The steady-state I–V relationship during the hyperpolarizing voltage steps shows inward rectification of that current. The peak amplitude at a voltage step of −100 mV was 233 ± 21 pA (mean ± SEM; $n = 7$) in isolated rods. Similar values were obtained from rods in slice preparations (247 ± 28 pA, $n = 6$).

The slow inward-rectifying currents and the inward tail currents were markedly reduced by 3 mM Cs^+, a blocker of a hyperpolarization-activated cation current [7, 10], in the bath, but not by 3 mM Ba^{2+} or 1 mM 4-acetamido-4'-isothiocyanostilbene-2,2'-disulfonic acid (SITS), a Cl^- current blocker [30]. This suggests that the inward currents during and after the command pulses are an h current (I_h). Bath application of 3 mM Cs^+ also blocked the steady inward current of approximately −20 pA at a V_h of −60 mV.

A concentration-response curve for block by Cs^+ at a voltage step of −100 mV could be fitted by the Hill equation with a half-blocking concentration (IC_{50}) of 41 μM and a Hill coefficient of 0.91.

When Cs^+ was added to the medium, the resting potential of the rod was hyperpolarized by approximately 10 mV, and the gradual decay of the voltage response was eliminated. The decay of the voltage response was also eliminated when the resting potential of the same rod was depolarized to the same membrane potential as in the control solution by injecting a steady positive current of +20 pA in the presence of Cs^+. These results suggest that the elimination of the voltage decay is caused by the Cs^+ block of I_h rather than by the change in the resting potential of rods.

The Kinetics of I_h

To distinguish the slow inward currents induced by hyperpolarization in human rods from anomalous rectifier K^+ currents reported in other preparations [31, 32], we examined the kinetics of the inward currents in detail. We found that the time course of the current activation was well described at all voltages by the sum of two exponentials by allowing the time constant and amplitude of the exponentials to change with voltage. The time constant of both exponentials increased linearly with voltage from an average value of 63 ms at −100 mV to 109 ms at −70 mV for a fast component and from 144 ms at −100 mV to 323 ms at −70 mV for a slow component. These values are similar to those of I_h in rods of cold-blooded vertebrates [9, 33], but much slower than those of anomalous rectifier K^+ currents in other tissues [31, 32]. This suggests that the slow inward currents induced by hyperpolarization in human rods are I_h rather than anomalous rectifier K^+ currents.

Modulation by I_h of Rod's Voltage Responses

To elucidate a mechanism underlying the modulation by I_h of voltage responses in human rods, we examined the effects of external Cs^+ on rod's voltage responses under current-clamp conditions. Isolated human rods that lacked an outer segment had a resting membrane potential of −54 ± 7 mV (mean ± SEM, $n = 5$) in the normal Ames' medium. Injecting a step of hyperpolarizing current mimicked the shape of the voltage response to light. Injection of −60 pA current step produced in rods a hyperpolarizing response that first reached a maximum and then gradually declined to a less polarized plateau.

This time course of decaying voltage responses was similar to that of I_h activation during hyperpolarizing voltage steps. Decaying voltage responses were well described at all voltages by the sum of two exponentials. The time constant of both exponentials increased linearly with peak voltage from an average value of 70 ms at -100 mV to 120 ms at -70 mV for a fast component and from 162 ms at -100 mV to 355 ms at -70 mV for a slow component.

Discussion

Voltage-Gated Na^+ Currents in Human Rods

We have shown that human rods express voltage-gated Na^+ channels, and can generate Na^+ action potentials. In contrast, photoreceptors of cold-blooded vertebrates such as toads [34], turtles [35, 36], and lizards [37] do not express voltage-gated Na^+ channels, but express voltage-gated Ca^{2+} channels and generate Ca^{2+} action potentials. In toad rods, depolarizing regenerative potentials occurred during the recovery phase of the light responses [34]. The amplitude of action potentials depends upon the extracellular Ca^{2+} concentration, and is increased by substitution of Sr^{2+} for Ca^{2+}. Action potentials in toad rods are blocked by 25 μM Cd^{2+} and 100 μM Co^{2+}, but unaffected by 1 μM TTX [34]. Similar results are reported in turtle cone photoreceptors [35, 36]. In addition, Maricq and Korenbrot [37] showed that, in addition to voltage-gated Ca^{2+} currents, Ca^{2+}-dependent Cl^- currents are also involved in spike generations in lizard cones. These observations suggest that action potentials, which occur in lower vertebrate photoreceptors, are not Na^+ spikes but Ca^{2+} spikes.

Comparison of Voltage-Gated Na^+ Currents with T-Type Ca^{2+} Currents

Among the several types of voltage-gated Ca^{2+} currents in various preparations, T-type Ca^{2+} currents are known to decay rapidly [27, 38, 39]. In the present study, the decay rate of the Na^+ currents evoked by membrane depolarization between -50 mV and $+10$ mV was 0.6–2.3 ms. These values are similar to the rates of voltage-gated Na^+ currents in other preparations (~0.5–2 ms; Kawai et al. [27], Hidaka and Ishida [29]), but much faster than those of T-type Ca^{2+} currents (~20–50 ms; Nowycky et al. [38], Kaneko et al. [39], Kawai et al. [27]). In the present study, however, we added 1 mM Co^{2+} to the bath solution in order to block synaptic inputs from other neurons; we cannot exclude the possibility that human rods may also express T-type Ca^{2+} channels.

Function of Na^+ Action Potentials in Human Rods

Na^+ action potentials in human rods may be generated normally in vivo, when their membrane potentials are hyperpolarized well below -50 mV by light. We suggest that Na^+ spikes, generated at the termination of an intensive and long light stimulus, may cause a greater than normal release of transmitter at that time by activating Ca^{2+} channels in the synaptic terminal, and thus selectively amplify the "off" part of the light signal. Furthermore, signaling via Na^+ spikes in human rods may serve to accelerate the termination of photovoltage signals, thereby securing the transmission of rapidly changing visual signals. Na^+ spikes in human rods might serve to accelerate the visual signals more rapidly compared with Ca^{2+} spikes in cold-blooded vertebrates.

Properties of I_h in Human Rods

Hyperpolarizing voltage steps from −60 mV induced slow inward-rectifying currents in human rods. This voltage dependence of the inward-rectifying currents is quite similar to that of I_h in photoreceptors of cold-blooded vertebrates [7–9] and monkey [10] retina and also similar to that of I_h in other preparations [40–42]. Bath application of Cs^+ reduced the amplitude of inward-rectifying currents in human rods in a dose-dependent manner. Hill coefficient for Cs^+ block of human inward-rectifying currents (0.91) was also similar to that in salamander rods. Wollmuth [43] reported that the inhibition curve of I_h in the salamander rods could be fitted by a Hill coefficient of 1.0 and showed that the blocking action of Cs^+ on I_h was noncompetitive. In contrast, 3 mM Ba^{2+}, an anomalous rectifier K^+ current blocker, or 1 mM SITS, a Cl^- current blocker, did not significantly change the inward-rectifying currents in human rods. A similar result was also reported in I_h of salamander rods [43].

The kinetics of inward-rectifying currents in human rods is voltage dependent and could be described as the sum of two exponentials with voltage-dependent time constants. The kinetic behavior of I_h in other photoreceptors has been also described by the sum of two exponentials in tiger salamander rods [33] and cones [9]. This kinetic scheme suggests the possibility that I_h channels exist in at least two closed and one open state [9, 33]. The time constants of activation of human inward-rectifying currents were similar to those of I_h in other preparations [9, 33], but much slower than those of anomalous rectifier K^+ currents in other tissues [31, 32], suggesting that the slow inward-rectifying currents in human rods are I_h rather than anomalous rectifier K^+ currents. As described earlier, the voltage dependence, pharmacology, and kinetics of human inward-rectifying currents suggest that I_h is expressed in human rod photoreceptors.

Physiological Role of I_h in Human Rods

The basic physiological function of I_h in various preparations is to depolarize cells after periods of hyperpolarization. In thalamic and cardiac cells I_h contributes to rhythmic firing by depolarizing the membrane potential following the hyperpolarization of an action potential [41, 42].

Under current-clamp conditions, voltage responses of human rods to depolarizing and hyperpolarizing current steps showed gradual decay. This suggests that the kinetics of voltage responses of human rods may be modulated by I_h. Bath application of 3 mM Cs^+ blocked the gradual decay in voltage responses. This suggests that I_h is responsible for the gradual decay in voltage responses of human rods.

In photoreceptors of cold-blooded vertebrates, I_h is thought to initiate the recovery from a strong hyperpolarization arising from bright flashes of light [7, 9, 17]. Fain et al. [17] reported that a bright flash produces in rods of cold-blooded vertebrates a hyperpolarizing response that first reaches a maximum and then gradually declines to a less polarized plateau. And they have suggested that the decay of the hyperpolarizing light response is produced by activation of I_h. Superfusing rods with a Cs^+-containing solution eliminated the gradual decay in the light response. These observations are also quite similar to voltage responses of human rods to injection of hyperpolarizing current steps.

Our study suggests that I_h in human rods becomes activated in the course of large hyperpolarizations generated by bright-light illumination, and therefore, I_h will generate a depolarizing component that opposes the light-generated hyperpolarization in human rods. The findings of the present study and the previous experiments from lower-vertebrate rods [7, 9, 17] may suggest that the mechanism underlying the modulation by I_h of rod's voltage responses is general in various vertebrates from lower vertebrates to humans. The recovery by I_h from a strong hyperpolarization arising from bright flashes of light may enhance temporal resolution in rod photoreceptors.

Acknowledgments. This work was done in collaboration with Dr. M. Horiguchi, Professor of the Department of Ophthalmology in Fujita Health University. We thank Drs. J. McReynolds and A. Kaneko for useful comments. This work was supported by Japan Society of the Promotion of Science and DAIKO FOUNDATION.

References

1. Lipton SA, Ostroy SE, Dowling JE (1977) Electrical and adaptive properties of rod photoreceptors in *Bufo marinus*. I. Effects of altered extracellular Ca^{2+} levels. J Gen Physiol 70:747–770
2. Copenhagen DR, Jahr CE (1989) Release of endogenous excitatory amino acids from turtle photoreceptors. Nature 341:536–539
3. Wässle H, Boycott BB (1991) Functional architecture of the mammalian retina. Physiol Rev 71:447–480
4. Masland RH (1996) Processing and encoding of visual information in the retina. Curr Opin Neurobiol 6:467–474
5. Savchenko A, Barnes S, Kramer RH (1997) Cyclic-nucleotide-gated channels mediate synaptic feedback by nitric oxide. Nature 390:694–698
6. DeVries SH, Schwartz EA (1999) Kainate receptors mediate synaptic transmission between cones and "Off" bipolar cells in a mammalian retina. Nature 397:157–160
7. Bader CR, Bertrand D, Schwartz EA (1982) Voltage-activated and calcium-activated currents studied in solitary rod inner segments from the salamander retina. J Physiol 331:253–284
8. Barnes S, Hille B (1989) Ionic channels of the inner segment of tiger salamander cone photoreceptors. J Gen Physiol 94:719–743
9. Maricq AV, Korenbrot JI (1990) Inward rectification in the inner segment of single retinal cone photoreceptors. J Neurophysiol 64:1917–1928
10. Yagi T, Macleish PR (1994) Ionic conductances of monkey solitary cone inner segments. J Neurophysiol 71:656–665
11. Schneeweis DM, Schnapf JL (1995) Photovoltage of rods and cones in the macaque retina. Science 268:1053–1056
12. Wollmuth LP, Hille B (1992) Ionic selectivity of I_h channels of rod photoreceptors in tiger salamanders. J Gen Physiol 100:749–765
13. Berry MJ, Brivanlou IH, Jordan TA, et al. (1999) Anticipation of moving stimuli by the retina. Nature 398:334–338
14. Schnapf JL, Nunn BJ, Meister M, et al. (1990) Visual transduction in cones of the monkey *Macaca fascicularis*. J Physiol 427:681–713
15. Schneeweis DM, Schnapf JL (1999) The photovoltage of macaque cone photoreceptors: adaptation, noise, and kinetics. J Neurosci 19:1203–1216
16. Schneeweis DM, Schnapf JL (2000) Noise and light adaptation in rods of the macaque monkey. Vis Neurosci 17:659–666

17. Fain GL, Quandt FN, Bastian BL, et al. (1978) Contribution of a caesium-sensitive conductance increase to the rod photoresponse. Nature 272:466–469
18. Kawai F, Horiguchi M, Suzuki H, et al. (2001) Na$^+$ action potentials in human photoreceptors. Neuron 30:451–458
19. Calkins DJ, Sterling P (1999) Evidence that circuits for spatial and color vision segregate at the first retinal synapse. Neuron 24:313–321
20. Pugh EN Jr, Nikonov S, Lamb TD (1999) Molecular mechanisms of vertebrate photoreceptor light adaptation. Curr Opin Neurobiol 9:410–418
21. Kawai F, Sterling P (1999) AMPA receptor activates a G-protein that suppresses a cGMP-gated current. J Neurosci 19:2954–2959
22. Werblin FS (1978) Transmission along and between rods in the tiger salamander retina. J Physiol 280:449–470
23. Hamill OP, Marty A, Neher E, et al. (1981) Improved patch-clamp techniques for high resolution current recording from cells and cell-free membrane patches. Pflügers Arch 391:85–100
24. Kawai F, Horiguchi M, Suzuki H, et al. (2002) Modulation by hyperpolarization-activated cation currents of voltage responses in human rods. Brain Res 943:48–55
25. Yau K-W, Baylor DA (1989) Cyclic GMP-activated conductance of retinal photoreceptor cells. Annu Rev Neurosci 12:289–327
26. Torre V, Ashmore JF, Lamb TD, et al. (1995) Transduction and adaptation in sensory receptor cells. J Neurosci 15:7757–7768
27. Kawai F, Kurahashi T, Kaneko A (1996) T type Ca^{2+} channel lowers the threshold of spike generation in the newt olfactory receptor cell. J Gen Physiol 108:525–535
28. Wang GY, Ratto G, Bisti S, et al. (1997) Functional development of intrinsic properties in ganglion cells of the mammalian retina. J Neurophysiol 78:2895–2903
29. Hidaka S, Ishida AT (1998) Voltage-gated Na$^+$ current availability after step- and spike-shaped conditioning depolarizations of retinal ganglion cells. Pflügers Arch 436:436–508
30. Tabata T, Ishida AT (1999) A zinc-dependent Cl$^-$ current in neuronal somata. J Neurosci 19:5195–5204
31. Hagiwara S, Kakahashi K (1974) The anomalous rectification and cation selectivity of the membrane of a starfish egg cell. J Membr Biol 18:61–80
32. Tachibana M (1983) Ionic currents of solitary horizontal cells isolated from goldfish retina. J Physiol 345:329–351
33. Hestrin S (1987) The properties and function of inward rectification in rod photoreceptors of the tiger salamander. J Physiol 390:319–333
34. Fain GL, Gerschenfeld HM, Quandt FN (1980) Calcium spikes in toad rods. J Physiol 303:495–513
35. Gerschenfeld HM, Piccolino M, Neyton J (1980) Feed-back modulation of cone synapses by L-horizontal cells of turtle retina. J Exp Biol 89:177–192
36. Piccolino M, Gerschenfeld HM (1980) Characteristics and ionic processes involved in feedback spikes of turtle cones. Proc R Soc Lond B Biol Sci 206:439–463
37. Maricq AV, Korenbrot JI (1988) Calcium and calcium-dependent chloride currents generate action potentials in solitary cone photoreceptors. Neuron 1:503–515
38. Nowycky MC, Fox AP, Tsien RW (1985) Three types of neuronal calcium channel with different calcium agonist sensitivity. Nature 316:440–443
39. Kaneko A, Pinto LH, Tachibana M (1989) Transient calcium current of retinal bipolar cells of the mouse. J Physiol 410:613–629
40. Mayer ML, Westbrook GL (1983) A voltage-clamp analysis of inward (anomalous) rectification in mouse spinal sensory ganglion neurones. J Physiol 340:19–45
41. DiFrancesco D (1984) Characterization of the pace-maker current kinetics in calf Purkinje fibres. J Physiol 348:341–367

42. McCormick DA, Pape HC (1990) Properties of a hyperpolarization-activated cation current and its role in rhythmic oscillation in thalamic relay neurones. J Physiol 431:291–318
43. Wollmuth LP (1995) Multiple ion binding sites in I_h channels of rod photoreceptors from tiger salamanders. Pflügers Arch 430:34–43

Neuroinformatics in Vision Science: NRV Project and VISIOME Platform

Shiro Usui

Key words. Neuroinformatics, Vision science, VISIOME, Mathematical model

Introduction

One of the frontiers of the twenty-first century is the elucidation of the complicated and elaborate functions of the brain, such as sensory perception, recognition, memory, emotion, etc. The specialization and segmentation of advanced research topics make it very difficult to integrate related evidence so as to understand the integrative functions of the brain. The introduction of information science technology in the analysis, processing, transmission, storage, integration, and utilization of information is indispensable.

Recalling the origins, in a recent report from the OECD Megascience Forum [1], the new field of neuroinformatics was defined as "the combination of neuroscience and information sciences to develop and apply advanced tools and approaches essential for a major advancement in understanding the structure and function of the brain." An emerging trend in modeling and understanding brain processes is that understanding the brain at one level can be greatly enhanced by considering the process embedded in its context. Equally important is a consideration of the complexities of neuronal processes operating in much detailed subsystems. A significant goal is to explore how abstractions at different levels are related, e.g., the pathways from molecular to system levels. Another critical goal is to discuss the disposition of the computational approach to support effective modeling across abstractions and subsystems. That is, neuroinformatics is a new research paradigm for the twenty-first century that fuses experimental techniques with mathematical and information science techniques. In particular, mathematical models are used to describe and integrate data and results obtained from a number of research fields. These mathematical models can be regarded as the platform that supports the simulation experiment,

Laboratory for Neuroinformatics, RIKEN BSI, Hirosawa, Wako, Saitama 351-0198; and Information and Computer Sciences, Toyohashi University of Technology, Tempaku, Toyohashi 441-8580, Japan

which is indispensable for studying and understanding the function and mechanism of the brain. In general, neuroinformatics should construct the support environment that integrates the databases devoted to various research fields with the data analysis techniques. If the research of each conventional field is the warp, neuroinformatics is the woof that links them.

Neuroinformatics Research in Vision (NRV) Project

The NRV project is the first project in Japan started in 1999 under the Strategic Promotion System for Brain Science of the Special Coordination Funds for Promoting Science and Technology at the Science and Technology Agency—now under the Ministry of Education, Culture, Sports, Science and Technology (MEXT)—aimed at building the foundation of neuroinformatics research. Because of the wealth of data on the visual system, the NRV project will use vision research to promote experimental, theoretical, and technical research in neuroinformatics [2].

The first goal of the project is to construct mathematical models for each level of visual system (single neuron, retinal neural circuit, visual function). The second goal is to build resources for neuroinformatics, utilizing information science technologies and the research support environment that integrates them. The third goal is to realize a new vision device based on the brain-type information processing principle. We have the following research groups and topics at present:

Group 1: Modeling a single neuron by mathematical reconstruction
Group 2: Realization of virtual retina based on cell physiology
Group 3: Study on the visual function by a computational and systems' approach
Group 4: Realization of artificial vision devices and utilization of silicon technology for recording and stimulation
Group 5: Development of fundamental neuroinformatics tools and environment

Details can be found at: http://www.neuroinformatics.gr.jp.

VISIOME Environment: Neuroinformatics Database and Environment for Vision Science

VISIOME's Enterprise Solution for Neuroinformatics approach to vision science can be stated as understanding that brain function requires the integration of diverse information from the level of the molecule to the level of neuronal networks. However, the huge amount of information is making it almost impossible for any individual researcher to construct an integrated view of the brain. To solve this problem, we are constructing a "Visiome Platform" as a test bed for a useful neuroinformatics environment. The basic concept is to make a web site integrating mathematical models, experimental data, and related information. The Visiome Platform has two major characteristic features. First, it will allow researchers to reuse the models and experimental data in the database. Researchers can see how the models work or compare their own results with other experimental data, improve or integrate models, and formulate their own hypothesis into a model. Second, it will provide a novel indexing system (Visiome Index) of the visual system research field, specifically oriented to the modeling studies. The Visiome Index is based on neuronal and cognitive functions that are important targets of modeling studies. That is, the Visiome Platform will provide not just a database of models and data but also powerful analysis libraries and simulation tools. All these features should be available on the platform including the Simulation Server and Personal Visiome.

Neuroinformatics Study on Vertebrate Retina as an Example

The vertebrate retina has been intensively studied over the past decades using physiological, morphological, and pharmacological techniques. The retina—thought to be a window to the brain because of its accessibility and suitability for the investigation— is a unique and interesting neural system from the viewpoint not only of neuroscience but also engineering. The retina is an ideal organ of the neural network system and information processing system. Since the early 1980s, enzymatic dissociation techniques have been employed for the isolation of a variety of retinal cells, including photoreceptors, horizontal cells, and bipolar cells. Using isolated cell preparations, membrane ionic currents of solitary retinal cells have been studied in a quantitative fashion using voltage-clamp techniques. These data provide information concerning the functional role of the ionic currents in generating and shaping the light response of retinal neurons. However, the detailed experimental data alone are not enough to understand how retinal neurons work. A combination of experimental work and mathematical modeling is necessary to clarify the complicated retinal functions. We believe that mathematical modeling and simulation analysis with an advanced computer system has the potential to help us understand the relationships between microscopic characteristics and their integrative functions of the neural system including the brain.

The purpose of our study is to develop a realistic model of the retina that replicates detailed neurophysiological structures and retinal functions. This approach allows the exploration of the computational functions performed by retinal neurons including the role and interaction of ionic channels and receptors, and the subcellular events such as transmitter release, binding, and uptake.

In the present study, mathematical descriptions of the membrane ionic currents in the retinal neurons, photoreceptors, horizontal cells, bipolar cells, and ganglion cells are realized. Electrical properties of each neuron are described by a parallel conductance circuit. The voltage- and time-dependent characteristics of ionic conductance are modeled by Hodgkin-Huxley types of equations. The developed model is capable of accurately reproducing the voltage- and current-clamp responses of retinal neurons. Any electrical response, including the light response that depends on the dynamic balance of different ionic currents, is quantitatively analyzed by the model. Therefore, hypotheses on how the retina processes visual information are understood at the cellular and subcellular level. Neuroinformatics models can be used to summarize what we know and what we need to find out on the retina [3]. In addition, such realistic models can be used to evaluate clinical effects and potential side effects of a particular drug, such as those induced by the bradycardic agent Zatebradine [4].

Conclusion

The NRV project is ongoing until the end of the 2003 fiscal year, and the prototype Visiome Platform will be open to the public soon. We welcome any comments and suggestions for establishing the platform in vision science.

Acknowledgments. The author thanks Drs. Masao Ito and Shun-ichi Amari at RIKEN BSI and the members of the NRV project for their support and collaboration.

References

1. Final report of the OECD Megascience Forum, Neuroinformatics: Working Group, June 2002, http://www.oecd.org/sti/gsf
2. Usui S, Amari S-I. (2002) Trends and scope of neuroinformatics research (in Japanese). Brain Science 24(1):11–17
3. Usui S, Kamiyama Y, Ishii H, et al. (1999) Ionic current model of the outer retinal cells. In: Poznanski RR (ed) Modeling in the neurosciences. Harwood Academic Publishers, Australia, pp 293–319
4. Usui S, Kamiyama Y, Ogura T, et al. (1996) Effects of Zatebradine (UL-FS 49) on the vertebrate retina. In: Toyama J, Hiraoka M, Kodama I (eds) Recent progress in electrophamacology of the heart. CRC Press, Boca Raton, pp 37–46

The Cone Pedicle, the First Synapse in the Retina

HEINZ WÄSSLE[1], SILKE HAVERKAMP[1], ULRIKE GRÜNERT[2], and
CATHERINE W. MORGANS[3]

Key words. Cone synapse, Synaptic ribbon, Glutamate receptor, Bipolar cell, Horizontal cell

Introduction

Cone pedicles are the output synapses of cone photoreceptors and transfer the light signal onto the dendrites of bipolar and horizontal cells. In the macaque monkey retina, cone pedicles contain between 20 and 45 ribbon synapses (triads), which are the release sites for glutamate, the cone transmitter [11, 13, 19, 22, 41]. Triads comprise a presynaptic ribbon, two horizontal dendrites as lateral elements, and one or two ON-cone bipolar cell dendrites as central elements (Fig. 1). Flat contacts are formed in most instances by the dendrites of OFF-cone bipolar cells at the cone pedicle base [19]. Glutamate released continuously at the synaptic ribbons acts not only through direct contacts but to an even greater extent through diffusion to the appropriate glutamate receptors (GluRs) [72].

Molecular cloning has revealed multiple types of GluRs [18, 30, 50], which are usually grouped into four major classes: AMPA receptors (subunits GluR1, GluR2, GluR3, and GluR4), kainate receptors (subunits GluR5, GluR6, GluR7, KA1, KA2, and possibly δ1 and δ2), NMDA receptors (subunits NR1, NR2A–D, NR3A), and metabotropic glutamate receptors (mGluR1–mGluR8).

Several studies have recently shown that horizontal and bipolar cells express different types of GluRs (for reviews see Brandstätter et al. [9]; Morigiwa and Vardi [46]; Qin and Pourcho [53]; Haverkamp et al. [27–29]; Hack et al. [26]; Brandstätter and Hack [7]. In the present study we concentrated on the localization of the AMPA receptor subunits GluR1, GluR2/3, and GluR4 and on the kainate receptor subunit GluR5.

Another transmitter involved with the cone pedicle synaptic complex is GABA, the putative horizontal cell transmitter [24]. GABA release from horizontal cells appears

[1] Max-Planck-Institute for Brain Research, Frankfurt, Germany
[2] Department of Physiology, University of Sydney, Sydney, Australia
[3] Neurological Science Institute, Oregon Health and Science University, Beaverton, USA

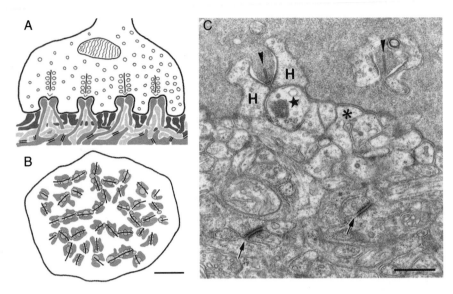

FIG. 1A–C. Structure of the cone pedicle of the macaque monkey retina. **A** Schematic drawing of side view through a cone pedicle. Four presynaptic ribbons, flanked by synaptic vesicles, and four triads are shown. Invaginating dendrites of horizontal cells (*middle grey*) form the lateral elements, and invaginating dendrites of ON-cone bipolar cells (*light grey*) form the central elements of the triads. OFF-cone bipolar cell dendrites (*dark grey*) make flat contacts at the cone pedicle base. Desmosome-like junctions (*small bar pairs*) are found at a distance of 1–2 μm *underneath* the pedicle. **B** Reconstruction of the horizontal, bottoms-up view of a cone pedicle [13]. The *short black lines* represent the ribbons. Invaginating ON-cone bipolar cell dendrites (*light grey*) and horizontal cell dendrites (*darker grey*) form the 40 triads associated with the ribbons. Flat contacts are not shown; however, they would cover most of the surface in the center of the pedicle. **C** Electron micrograph of a vertical section through the synaptic complex of a cone pedicle base. The *upper third* of the micrograph shows the cone pedicle filled with vesicles. A typical triad, containing a presynaptic ribbon (*arrowhead*), two lateral horizontal cell processes (*H*), and one invaginating bipolar cell dendrite (*star*) is present in the *upper left part*. A flat (basal) contact is indicated by an *asterisk*. Two desmosome-like junctions are indicated by *arrows*. **B** *Bar* 2 μm; **C** *Bar* 0.5 μm

to be Ca^{2+} independent, and it has been proposed that a GABA transporter mediates this release [68, 69]. Dendrites contacting the cone pedicles are, therefore, exposed to both glutamate and GABA released in a light-dependent manner.

Vertical Organization of Glutamate Receptors at the Cone Pedicle Base

Cones of the macaque monkey retina are arrayed in a regular mosaic (Fig. 2B), and the cone pedicles are lined up along the outer margin of the outer plexiform layer (OPL; Fig. 2A). Glutamate receptors in the OPL are aggregated in groups of puncta (hot spots) that are in register with the cone pedicles (Fig. 2C–F). In the case of GluR1 (Fig. 2C) and GluR5 (Fig. 2F), one band of puncta can be detected; in the case of GluR2/3 (Fig. 2D) and GluR4 (Fig. 2E), two bands, 1.5 μm apart, are aligned with the cone pedicle base.

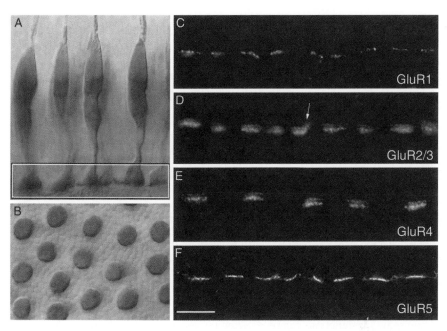

Fig. 2A–F. Localization of glutamate receptor (*GluR*) subunits at the macaque monkey cone pedicle base. A Vertical view of four cones that were immunostained for calbindin; the *box* outlines the cone pedicle region and all the following fluorescence micrographs illustrate this small part of the retina. B Horizontal view of the cones (calbindin-immunolabeled) and rods (unlabeled) in the peripheral retina. C–F Fluorescence micrographs of vertical cryostat sections through the outer plexiform layer (OPL) that were immunolabeled for GluR1, GluR2/3, GluR4, and GluR5. Immunofluorescence is punctate and confined to individual cone pedicles (nine in C, nine in D, five in E, and eight in F). The *arrow* in D points to weak label in rod spherules. Bars 10 μm

To define the relative position of the bands with respect to the cone pedicle base, we double-labeled sections for GluRs and postsynaptic density protein 95 (PSD-95). PSD-95 has been shown previously to label the presynaptic membrane of cone pedicles and rod spherules [37]. Figure 3A–C shows that the GluR1 immunoreactive puncta coincide with the basal membrane of cone pedicles. The same holds true for the GluR5 band (not shown). The two GluR4 bands (Fig. 3D–F) showed an unexpected stratification: the upper band coincided, in fact, with the cone pedicle base, while the lower band was 1.5 μm below the cone pedicle base. The same holds true for the GluR2/3 band (not shown).

Desmosome-like Junctions Represent Glutamate Receptor Clusters

To further analyze this surprising clustering of GluRs below the cone pedicle base, electron microscopy was applied to well-fixed vertical sections (Fig. 1C). The known structure of the cone pedicle, with ribbons and invaginating and flat contacts, is apparent in Fig. 1C; in addition, at a distance of 1.5 μm from the cone pedicle base,

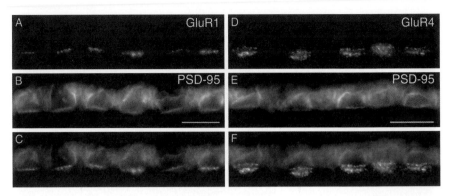

FIG. 3A–F. Localization of GluR1 and GluR4 at the cone pedicle base. **A–C** Fluorescence micrographs of a vertical cryostat section through the OPL that was double labeled for GluR1 (**A**) and postsynaptic density protein 95 (*PSD-95*); **B** The outer membranes of cone pedicles (six larger profiles) and rod spherules (smaller, round profiles) are strongly immunoreactive for PSD-95. The micrograph in **C** shows a superposition of **A** and **B**. **D–F** Fluorescence micrographs of a vertical cryostat section through the OPL that was double labeled for GluR4 (**D**) and PSD-95 (**E**). The micrograph in **F** shows their superposition. *Bars* 10 μm

desmosome-like junctions can be detected (arrows in Fig. 1C). Although these junctions had been observed in the earliest electron micrographs of cone pedicles and were also described in a freeze fracture study [57], their true nature is not understood [5]. We immunostained monkey retinae for proteins that are usually expressed in adherens junctions, desmosomes, and tight junctions [25]; none labeled the desmosome-like junctions in the OPL [27].

Next we tested the hypothesis that these desmosome-like junctions correspond to the GluR2/3- and the GluR4 immunoreactive hot spots below the cone pedicle. Electron micrographs of GluR1 and GluR4 immunolabeling are compared in Fig. 4A,B. The GluR1 subunit was generally found in putative OFF-cone bipolar cells at basal junctions, but not at desmosome-like junctions (Fig. 4A). A comparable clustering at bipolar cell basal junctions was also observed in the case of the GluR5 subunit [29]. GluR4 immunolabeling was observed at basal junctions, in invaginating horizontal cell processes (Fig. 4B; one or both were found to be labeled), and in lateral elements of rod spherules (not shown). In addition, GluR4 immunoreactivity was also observed at the desmosome-like junctions below the cone pedicle (arrows in Fig. 4B).

Several proteins have been shown recently to be involved with the clustering and anchoring of transmitter receptors at postsynaptic densities [21, 36]. We observed one of them, SAP102 (synapse associated protein 102), both at the invaginating processes of horizontal cells and at the desmosome-like junctions (not shown). All this taken together suggests that desmosome-like junctions are not desmosomes but rather postsynaptic densities at which GluRs appear to be clustered.

Horizontal Cell Processes Form Desmosome-like Junctions

To find out which cell types express GluR4 at the desmosome-like junctions, we used light microscopy and double-immunolabeling techniques. Parvalbumin (PV) in the

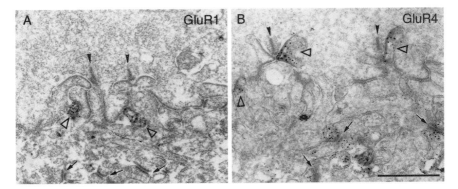

Fig. 4A,B. Electron micrographs of the synaptic complex at the cone pedicle base. In this and the following electron micrographs, the immunolabeling tolerated only short fixation times; therefore, the tissue preservation is compromised. **A** The *upper part* shows a cone pedicle and two synaptic ribbons (*filled arrowheads*). Two flat contacts of bipolar cells (*open arrowheads*) express GluR1 immunoreactivity. Several desmosome-like junctions (*arrows*) appear unlabeled. **B** GluR4 label is present in two invaginating horizontal cell dendrites and one flat bipolar cell contact (*open arrowheads*). Several desmosome-like junctions (*arrows*) show GluR4 immunoreactivity. *Bar* 1 μm

OPL is a selective marker of both H1 and H2 horizontal cells [59, 80]. In a section that was double labeled for PV and GluR4 (Fig. 5A–C), a close correspondence between the dendritic tips of horizontal cells at the cone pedicles and the GluR4 immunoreactive puncta can be detected. The gap between the two bands of GluR4 immunoreactive puncta (Fig. 5A) is also present in the PV immunoreactive dendritic tips of horizontal cells (Fig. 5B).

We also studied PV-immunolabeled sections by electron microscopy. The section in Fig. 5D shows the neuropil underneath a cone pedicle base. Invaginating horizontal cells that form the lateral elements of triads are labeled. A band of 1–2 μm width underneath the cone pedicle contains only a few labeled horizontal cell processes; beyond that gap, PV immunoreactive horizontal cell processes are engaged in contacts at the desmosome-like junctions. Among 26 junctions, we found that in 84% both sides were labeled, in 8% only one side was labeled, and in 8% both sides were unlabeled. The great majority of desmosome-like junctions are, therefore, between adjacent horizontal cell processes. However, we cannot exclude that there are also bipolar cell dendrites involved with these junctions.

Taken together, the results reveal a very precise vertical organization of GluRs underneath the cone pedicle and show that (1) the glutamate receptor subunits GluR1 and GluR5 are aggregated at the flat contacts made by putative OFF-cone bipolar cells at the cone pedicle base; (2) the subunits GluR2/3 and GluR4 are aggregated at the invaginating dendrites of horizontal cells, at some flat contacts at the cone pedicle base, and at the desmosome-like junctions; and (3) altogether three distinct layers of GluR clusters can be distinguished [27–29].

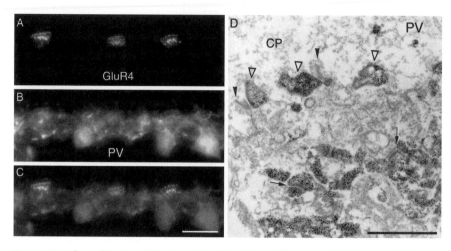

Fig. 5A–D. The architecture of horizontal cell dendrites at the cone pedicle base. A–C Fluorescence micrographs of a vertical section through the OPL that was double labeled for GluR4 (A) and parvalbumin (PV, B). The section passed through three cone pedicles. C Superposition of A and B shows the close correspondence between GluR4 clusters and the dendritic tips of horizontal cells. D Electron micrograph of a vertical section through the cone pedicle base (CP) that was immunolabeled for PV. PV immunoreactivity is prominent in the dendritic tips of horizontal cell dendrites inserted as lateral elements into the triads (open arrowheads). PV immunoreactivity is also found in the horizontal cell processes that form the desmosome-like junctions (arrows), approximately 1.5 μm underneath the cone pedicle base. Filled arrowheads point to the ribbons. A–C Bars 10 μm; D Bar 1 μm

Localization of the Presynaptic Cytomatrix Protein Bassoon at the Cone Pedicle Base

The precise role of the synaptic ribbons in directing synaptic vesicles to the active zone where the vesicles fuse with the membrane and release glutamate still needs to be elucidated [22, 43, 44, 55]. However, all the evidence available suggests that glutamate release from the cone pedicle occurs at the active zones below the ribbons. The relative position of postsynaptic GluRs with respect to the ribbons is therefore important, because it defines the speed and reliability of synaptic transmission.

Recently our lab has shown that the presynaptic cytomatrix protein bassoon [73] is associated with the synaptic ribbons of cone pedicles and rod spherules [10]. The electron micrograph in Fig. 6A shows that bassoon immunoreactivity is restricted to the active zone and the ribbon base both in cone pedicles and rod spherules. This restricted distribution of bassoon offers the possibility to study the precise arrangement of ribbons in rod spherules and cone pedicles by confocal light microscopy and thus avoid the difficult and time-consuming EM reconstructions. The inset in Fig. 6A shows the bassoon-labeled ribbons in a vertical section through the OPL: they are horseshoe-shaped in rod spherules and form strings along the cone pedicle base (arrowheads).

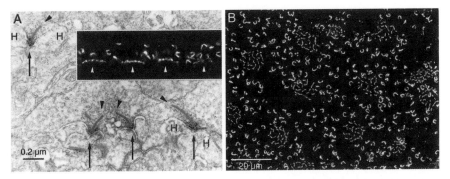

FIG. 6A,B. Localization of the presynaptic cytomatrix protein bassoon at the ribbons of rod spherules and cone pedicles. **A** Electron micrograph of a vertical section through a rod spherule (*top left*) and a cone pedicle (*bottom*) of the rat retina (kindly provided by J.H. Brandstätter). The *arrowheads* point to the presynaptic ribbons and the *arrows* indicate bassoon immunoreactivity, which decorates the bottom parts of the ribbons close to the arcuate densities. The *inset* in **A** shows a fluorescence light micrograph of a vertical section through four cone pedicles (*arrowheads*) and several rod spherules of the macaque monkey retina. **B** Confocal horizontal section through the cone pedicles (clusters of small ribbons) and rod spherules (horseshoe-shaped ribbons) of peripheral macaque monkey retina. The retina was immunostained for bassoon

The arrangement of ribbons within individual cone pedicles becomes more apparent in the horizontal view of a retinal whole mount (Fig. 6B). Between 20 (fovea) and 45 (peripheral retina) ribbons are contained within one cone pedicle. Their distribution within the pedicle appears irregular and resembles the pattern derived from EM reconstructions (see Fig. 1B). Recently, further immunocytochemical markers of ribbons have been described, resulting in comparable patterns of distribution (kinesin motor KIF3A [47]; RIBEYE [65]).

Localization of the α1F Calcium Channel Subunit at the Cone Pedicle

The electrophysiological properties of calcium channels and their distribution at the cone pedicle are likely to be the key determinants of the release properties at this synapse. The calcium channels of fish and amphibian photoreceptors are sensitive to dihydropyridines (DHPs); thus, they have been classified as L-type [58,64,81]). Localization of the α1C calcium channel subunits to the synaptic layers of the tiger salamander retina suggests that L-type channels mediate transmitter release in the amphibian retina [48]. The calcium currents of cones in the tree shrew, and monkey retinas, have been characterized electrophysiologically and appear similar to L-type currents [71,84]. Recently, a novel calcium channel gene, *CACNA1F*, has been identified that encodes the α1 subunit of a retina-specific, voltage-gated calcium channel, α1F [4,70]. Sequence comparisons show that α1F is a member of the L-type family of α1 subunits. In the rat retina, α1F immunoreactivity has been localized to photoreceptor cell bodies, and rod spherules and cone pedicles [45]. Mutations in

CACNA1F cause incomplete X-linked congenital stationary night blindness (CSNB2), a recessive nonprogressive visual disease in humans, the phenotype of which is consistent with a defect in neurotransmission between photoreceptors and second-order neurons [42]. Therefore, it is important to study the precise localization of the α1F subunit also in the primate retina.

Figure 7A–C shows confocal horizontal sections through a whole mount of a macaque monkey retina that was double immunolabeled for bassoon (Fig. 7A) and for the α1F subunit (Fig. 7B). Bassoon immunofluorescence reveals the ribbons of one single cone pedicle in the center (Fig. 7A), surrounded by the horseshoe-shaped ribbons of many rod spherules. The α1F subunit (Fig. 7B) appears to be restricted to the synaptic ribbons of the cone pedicle in the center and the surrounding rod spherules. This is confirmed by the superposition of the bassoon and α1F immunoreactivities in Fig. 7C. Close inspection of the horseshoe-shaped ribbons in Fig. 7C shows that the red α1F signal is slightly displaced towards the center of the horseshoe. This light microscopic analysis suggests that the α1F subunit is concentrated at the active zone underneath the ribbon where the synaptic vesicles undergo exocytosis.

This very precise localization of the α1F subunit at the ribbons supports the notion that the release of glutamate at cone pedicles is focal and restricted to the active zone of the triads [55]. Depending on the relative position of the postsynaptic contacts with respect to the active zone, a different efficacy of the released glutamate can be expected. Thus, we also studied the precise localization of the postsynaptic GluR clusters with respect to the synaptic ribbons.

Localization of the GluR Subunits GluR1, GluR4, and GluR5 at the Cone Pedicle Base

Figure 7D–F shows a horizontal confocal section of monkey retina that was double labeled for bassoon and the GluR4 subunit. The focal plane was at the outer band of GluR4 immunoreactive hot spots and does not show GluR4 clusters at the desmosome-like junctions. The cone pedicle contains 36 ribbons (Fig. 7D; the arrowhead indicates a rod spherule) and 39 GluR4 immunoreactive hot spots (Fig. 7E). Superposition of the two micrographs in Fig. 7F shows that the great majority of the GluR4 immunoreactive puncta (90%) are in register with ribbons, and only a few puncta (10%) are not associated with the ribbons. The GluR4 clusters at the desmosome-like junctions were not in register with the ribbons (not shown). Comparable double-labeling experiments were also performed for the GluR2/3 subunits and the puncta in the outer band were also found in register with the ribbons (not shown). A more quantitative evaluation of a total of 24 cone pedicles showed that (1) there are always slightly more GluR2/3 and GluR4 puncta than there are ribbons, (2) 88% of these puncta are in register with the ribbons, (3) only 12% do not coincide with the ribbon, and (4) every individual ribbon is in register with a GluR2/3 and GluR4 immunoreactive hot spot.

The situation was different in horizontal sections that were double labeled for bassoon and GluR1 (Fig. 7G–I). Electron microscopy (Fig. 4A) predicts that the GluR1 subunit is only found at flat contacts formed by OFF-cone bipolar cells at the cone pedicle base.

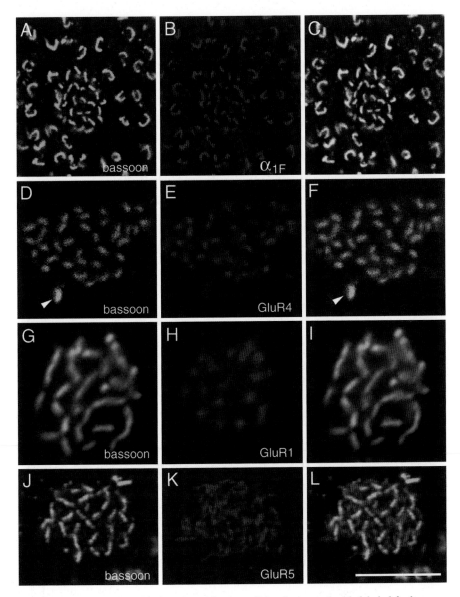

Fig. 7A–L. Horizontal confocal sections of cone pedicles that were double labeled for bassoon (*green*) and other synaptic markers (*red*). A The ribbons of one cone pedicle in the center and many ribbons of rod spherules (horseshoe-shaped) are immunolabeled for bassoon. B Same section as in A but immunostained for the calcium channel subunit α1F. C Superposition of A and B shows that the α1F subunit is aggregated at the ribbons of cone pedicles and rod spherules. D The ribbons of one cone pedicle and one rod spherule (*arrowhead*) are immunoreactive for bassoon. E GluR4 immunoreactive hot spots of the same section as in D. F Superposition of D and E shows that most of the GluR4 immunoreactive hot spots are in register with the ribbons. G Cone pedicle from the more central retina immunolabeled for bassoon. H GluR1 immunoreactive hot spots of the same section as in G. I, Superposition of G and H shows that only half of the GluR1 labeled hot spots are in register with the ribbons, and the other half are displaced from the ribbons. J Cone pedicle immunolabeled for bassoon. K Same pedicle, immunolabeled for GluR5. L Superposition of J and K shows that GluR5 immunoreactive hot spots (*red*) are found in between the ribbons (*green*). A–C *Bars* 15 μm; D–L *Bars* 5 μm

Reconstructions of Golgi-stained OFF bipolar cell cone contacts by EM [5, 31] have shown that flat contacts can be separated into those close to the triads (triad associated; TA) and those farther away from the triads (nontriad associated; NTA). The TA flat contacts did not directly coincide with the ribbons but were slightly displaced from the ribbons by the invaginating processes.

The cone pedicle in Fig. 7G contains 30 ribbons and 20 GluR1 immunoreactive hot spots (Fig. 7H). The superposition of bassoon and GluR1 labeling in Fig. 7I shows that some GluR1 immunoreactive puncta are in close association with the ribbons, whereas others are not. However, even those GluR1 immunoreactive puncta (red) that are associated with ribbons are not precisely in register, compared with the GluR4 puncta in Fig. 7F. It is, therefore, possible in the confocal micrographs to classify flat contacts into TA and NTA contacts. A more quantitative evaluation of a total of 11 cone pedicles showed that (1) there are always more ribbons ($n = 36.3 \pm 5$) than GluR1 immunoreactive puncta ($n = 25 \pm 5$), (2) 52% of these puncta are TA, (3) 48% are NTA, and (4) 65% of the ribbons are not associated with GluR1 immunoreactive hot spots.

We also studied by confocal light microscopy GluR5 clusters and their relative position with respect to the ribbons (Fig. 7J–L). The superposition of bassoon and GluR5 labeling in Fig. 7L shows that most of the GluR5 clusters occupy the space in between the ribbons, and only few appear to be associated with the ribbons. This finding suggests that most flat contacts, where GluR5 is expressed, are NTA contacts. The number of ribbons and the number of clusters were counted in seven cone pedicles. The number of GluR5 clusters (81.9 ± 6) was higher than the number of ribbons (42.6 ± 5) by a factor of 1.9. We also performed a double-labeling experiment for GluR1 and GluR5 and did not observe any colocalizations of the respective clusters (not shown).

Localization of GABA Receptors at the Cone Pedicle Base

There are several reports with respect to the localization of $GABA_A$ receptors ($GABA_A$ Rs) at the cone pedicle base. Vardi et al. [76] described in the cat retina that cone pedicles and bipolar cell dendrites express the $\alpha 1$ subunit of the $GABA_A$ R. In the primate retina, label was restricted to bipolar cell dendrites [75, 78]. In the rabbit retina, $GABA_A$ R $\alpha 1$ immunoreactivity was present on dendrites of cone bipolar cells adjacent to the cone pedicles [23]. $GABA_C$ receptor ρ subunits have also been described in the OPL [20]. After observing the laminated architecture of horizontal cell and bipolar cell dendrites underneath the cone pedicles, we sought to study the fine architecture of GABA Rs within this network of processes.

Figure 8A–C shows a section through five cone pedicles that were double labeled for GluR4 and the $\alpha 1$ subunit of the $GABA_A$ R. GluR4 immunoreactivity (Fig. 8A) shows the two bands of hot spots, while the $GABA_A$ R $\alpha 1$ subunit is concentrated in a single band (Fig. 8B). The superposition of the two micrographs (Fig. 8C) demonstrates that $GABA_A$ Rs are sandwiched neatly between the upper and lower bands of GluR4 immunoreactivity. Weak $GABA_A$ R immunofluorescence is also present along the ascending dendrites of bipolar cells and at their parent cell bodies (not shown).

Figure 8D–F shows a section that was double labeled for GluR2 (Fig. 8D) and the ρ subunits of the $GABA_C$ R (Fig. 8E). GluR2 immunoreactivity is concentrated at five cone pedicles in two bands of hot spots. $GABA_C$ R expression is prominent at these

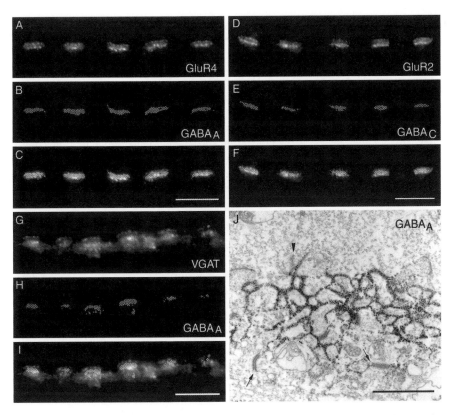

FIG. 8A–J. Localization of the α1 subunit of the (*GABA*)$_A$ receptor (GABA$_A$ R) and of the ρ subunits of the GABA$_C$ R in the OPL. **A–C** Composite fluorescence micrographs of a vertical section, through five cone pedicles, that was double labeled for GluR4 (**A**) and for the α1 subunit of the GABA$_A$ R (**B**). In this and the following micrographs, GABAR immunofluorescence is represented by the *stippled region*. **C** The superposition of **A** and **B** shows that the GABA$_A$ Rs (*stippled region*) are sandwiched between the two GluR4 bands. **D–F** Composite fluorescence micrographs of a vertical section, through five cone pedicles, that was double labeled for GluR2 (**D**) and the ρ subunits of the GABA$_C$ R (**E**). **F** The superposition of **D** and **E** shows that the GABA$_C$ Rs (*stippled region*) are sandwiched between the two GluR2 bands. **G–I** Fluorescence micrographs of a slightly oblique section, through seven cone pedicles, that was double labeled for the vesicular GABA transporter (*VGAT*; **G**) and the α1 subunit of the GABA$_A$ R (**H**). **I** The superposition of **G** and **H** shows that the GABA$_A$ Rs (*stippled region*) are sandwiched between the two bands of most intense VGAT expression. **J** Electron micrograph of a section through the cone pedicle synaptic complex that was immunolabeled for the α1 subunit of the GABA$_A$ R. GABA$_A$ Rs decorate the dendrites of bipolar cells in a narrow band in between the cone pedicle base and the desmosome-like junctions (*arrows*). The *filled arrowhead* points to a ribbon. **A–I** *Bars* 10 μm; **J** *Bar* 1 μm

five cone pedicles. Bipolar cells and their ascending dendrites are weakly labeled (not shown). The superposition of GluR2 and $GABA_C$ immunostaining (Fig. 8F) shows that $GABA_C$ Rs are sandwiched in between the GluR2 immunoreactive bands.

We studied the localization of $GABA_A$ Rs also by electron microscopy. The section in Fig. 8J passes through a cone pedicle, and $GABA_A$ R immunoreactivity decorates the membranes of bipolar cell dendrites in between the cone pedicle base and the desmosome-like junctions. Both the light and electron microscopic results of Fig. 8 show that the GABA Rs are concentrated underneath the cone pedicle; however, they are not aggregated in synaptic hot spots. We would argue that this more distributed localization is adapted to the mode of action of GABA in the OPL. Apart from some few interplexiform cell processes, GABA in the OPL is released principally from the horizontal cells and this release appears to be mediated by a GABA transporter [68, 69].

GABA Release from Horizontal Cells

Staining of mammalian retinae for the GABA plasma membrane transporters GAT-1, GAT-2, and GAT-3 did not show any labeling of horizontal cells and it is, therefore, unlikely that GAT-1-3 represent the horizontal cell GABA transporter [32–34]. We repeated the immunostaining of GAT-1 and GAT-3 in the monkey retina and did not observe any labeling of horizontal cells. In contrast, amacrine cells were intensely labeled (not shown). We did not test GAT-2 since it was expressed only in cells of the pigment epithelium of the rat retina [34]. Recently, antibodies that recognize the vesicular GABA transporter (VGAT) became available [40, 60] and were found in the rodent retina to label amacrine and horizontal cells [14]. We applied these antibodies to sections of the monkey retina and performed a double labeling for the VGAT (Fig. 8G) and for the $GABA_A$ R $\alpha 1$ subunit (Fig. 8H). VGAT immunoreactivity in the OPL is strongest at the cone pedicle base (Fig. 8G) and weak label also decorates the horizontal cell perikarya (not shown). The highest expression of VGAT appears to coincide with the invaginating contacts of horizontal cells and the layer of the desmosome-like junctions 1–2 μm underneath the cone pedicles. Because the section in Fig. 8G is slightly oblique, the cone pedicle base appears more blurred than in the other micrographs. Expression of the $GABA_A$ R $\alpha 1$ subunit is concentrated underneath the cone pedicle (Fig. 8B,H). Superposition of VGAT and the $GABA_A$ R $\alpha 1$ subunit shows that the receptors are sandwiched in between the two bands of high VGAT expression (Fig. 8I).

We do not yet know whether VGAT, which is apparently aggregated in the dendritic terminals of horizontal cells, is associated with synaptic vesicles, or whether it is inserted into the plasma membrane. The first case would predict a vesicular release of GABA, while the second case would be in favor of the transporter-mediated release. Leaving aside for the moment the precise mode of GABA release, the concentration of VGAT at the dendritic tips suggests that GABA is locally released there. And more important, there seems to be a good spatial match between the GABA Rs expressed along the bipolar cell dendritic terminals and the expression of VGAT in horizontal cell dendrites.

Discussion

Cone pedicles, the output synapses of cone photoreceptors, transfer the light signal onto the dendrites of bipolar and horizontal cells. Cone pedicles of the macaque monkey retina contain between 20 (fovea) and 45 (peripheral retina) ribbons. Synaptic vesicles are aggregated at the ribbons and we have shown here that the $\alpha 1F$ calcium channel subunit is concentrated at the active zone at the base of the ribbon [45]. Most components of the molecular machinery governing transmitter release at conventional synapses are also aggregated along the ribbon and at the active zone (for review, see Morgans [43,44]). Taken together, this suggests that the calcium-dependent release of glutamate occurs at the active zone underneath the ribbon.

The postsynaptic elements of the cone pedicle have until now been thought to consist of the invaginating dendrites of horizontal and bipolar cells at the triads and the basal contacts of bipolar cells [19, 57] (Fig. 1). In the present study we add a third postsynaptic structure, the desmosome-like junctions underneath the cone pedicle. It is likely that glutamate released at the ribbons would have to diffuse to these three postsynaptic structures to exert its function. One might argue that the desmosome-like junctions are too far away from the release sites and action of glutamate by diffusion is not possible. However, Müller cell processes, which perform the most efficient glutamate uptake in the retina, are absent from cone pedicles and rod spherules [16, 52, 56, 62]. Theoretical considerations have shown that such an action by diffusion is possible [55, Ribble et al. 1997, Invest Ophthalmol Vis Sci., abstract]. The diffusion of glutamate from the release sites at the ribbons to the desmosome-like junctions would take approximately 1 ms [3], which is an order of magnitude smaller than the delay between the absorption of a photon and the electrical response of the cone [66, 67]. Moreover, an anatomical and physiological study of the cone synapse of the goldfish retina has shown that the diffusion of glutamate is much faster than the horizontal cell dynamics [74]. This diffusion of glutamate contributes, therefore, only a minute fraction to the total delay of the retinal light response.

It is interesting in this context that recently in the cerebellar granular layer similar junctions were observed, named attachment plaques, and that NMDA receptors were aggregated at these plaques [51]. Hence, they might be a more general structure in the nervous system.

Glutamate Receptors of Horizontal Cells

In the macaque monkey retina, two types of horizontal cells, H1 and H2 cells, have been described (for review, see Wässle et al. [80]). Since two lateral horizontal cell processes are inserted into every triad, the total number of horizontal cell contacts of a peripheral cone pedicle is about 100 [13]. It is an attractive idea to postulate that one of the two lateral elements originates from an H1 cell and the other from an H2 cell, and moreover that they express different subsets of glutamate receptors. Recently it became possible to study patches of peripheral monkey retina, where all H2 horizontal cells and their cone contacts were stained by the injection of the tracer neurobiotin [15, 80]. The L/M (red/green) cone pedicles in such patches receive an average of 6.3 ± 1.9 ($n = 30$) invaginating processes from H2 horizontal cells. Cone pedicles of peripheral retina contain 45 triads; hence, an H2 horizontal cell process is only present in 6 or 7 of the 45 triads. In 80% of the triads, both lateral elements are likely to originate from H1 horizontal cells.

Extrasynaptic labeling of horizontal cells in the cat retina showed that the GluR4 subunit is only expressed by A-type horizontal cells, whereas the GluR 2/3 subunit is expressed by both A- and B-type horizontal cells [46, 53]. H2 cells of the primate retina are homologous to cat A-type horizontal cells [61, 80]. H2 cells preferentially innervate S (blue)-cone pedicles [12] and if, as in the cat retina, they specifically express the GluR4 subunit, S-cone pedicles should have a preponderance of GluR4 puncta. However, this does not seem to be the case because in *all* cone pedicles observed, we found, comparable to Fig. 7D–F, GluR4 hot spots at every bassoon labeled ribbon. We also observed (not shown) GluR2- and GluR2/3-labeled clusters at every bassoon labeled ribbon. Our EM observations [Haverkamp et al. 2000a,b, 2001] show that both horizontal cell processes inserted into the triads can express the same GluR subunit.

All these results suggest that there may be no difference in the expression of the AMPA receptor subunits GluR2, GluR2/3, and GluR4 (1) between the two horizontal cell processes inserted into the triads, (2) between H1 and H2 horizontal cells, and (3) between the different cone types. However, although these AMPA receptor subunits appear to take a major share in the transfer of the light signal from cones to horizontal cells, we also observed the expression of the kainate receptor subunit GluR6/7 on a few invaginating horizontal cell dendrites and at the desmosome-like junctions [29]. This expression was only found at M/L cone pedicles and was absent from S-cone pedicles, and we interpret this result as a difference between H1 and H2 cells: both H1 and H2 cells receive their cone input mainly through AMPA receptors, but H1 cells in addition express the kainate receptor subunit GluR6/7.

Glutamate Receptors of Bipolar Cells

In the macaque monkey retina, five different types of putative ON-cone bipolar cells have been described [6]. The diffuse bipolar cells DB4, DB5, and DB6 contact several neighboring cones, the invaginating midget bipolar cell (IMB) contacts a single cone pedicle, and the blue cone selective (BB) bipolar cell connects to several S-cone pedicles. The total number of ON bipolar cell contacts of a peripheral L/M cone pedicle is about 100 [13]. We did not study in the present paper the glutamate receptors expressed at the invaginating contacts of ON-cone bipolar cells. In the rodent retina it has been shown by immunocytochemistry that the metabotropic glutamate receptor mGluR6 is expressed at the dendritic tips of ON bipolar cells [49, 78]. In the primate retina, mGluR6 has also been localized to the invaginating ON bipolar cell dendrites [79]. However, because putative ON bipolar cells of the primate retina also make flat contacts at the cone pedicle base [31], mGluR6 was also observed at some flat contacts [79]. In addition to mGluR6, other metabotropic GluRs were observed at ON bipolar cell dendrites [9] and it can be expected that ON bipolar cell responses to glutamate are more diverse than assumed so far [2, 83].

In the macaque monkey, four different types of putative OFF-cone bipolar cells have been described [6]. The diffuse bipolar cells DB1, DB2, and DB3 connect, exclusively through flat contacts, to several neighboring cone pedicles. The flat midget bipolar cell (FMB) contacts a single cone pedicle [31]. The estimated number of OFF bipolar cell flat contacts of a peripheral L/M cone pedicle is 200–300 [13, 31]. We have shown here that the AMPA receptor subunit GluR1 is exclusively localized at flat contacts. We have also observed a small number of flat contacts immunoreactive for the AMPA receptor GluR2, GluR2/3, and GluR4 subunits [28]. However, adding up all AMPA receptor clusters results in only a small fraction of the total number of GluR clusters needed to serve all 200–300 OFF-cone bipolar cell contacts. We have also shown that the kainate receptor subunit GluR5 provides approximately 81 clusters per cone pedicle base. In a previous study, it was demonstrated that about 46 hot spots express the kainate receptor GluR6/7 subunit [29]. Brandstätter et al. [8] described in the rat retina many flat contacts expressing the kainate receptor KAII subunit. All this suggests that kainate receptors are the backbone for the signal transfer between cones and OFF-cone bipolar cells.

Using electrophysiological means, DeVries [17] has recently characterized the glutamate receptors expressed at flat contacts in the ground squirrel retina. He identified one type of OFF-cone bipolar cell, the b2 cell, that received input from the cone pedicles through AMPA receptors. Two other types, b3 and b7, received their light signals through kainate receptors. He went on to demonstrate that the signal transfer between cones and OFF-cone bipolar cells was faster in the case of b2 cells (S.H. DeVries, personal communication). He concludes that this is not caused by a difference in the molecular properties of the GluRs but by a closer apposition of b2 contacts to the ribbons. This would be in agreement with our anatomical findings (Fig. 7) showing that GluR1 hot spots are triad associated, while GluR5 hot spots are non-triad associated.

The Release of GABA from Horizontal Cells

Horizontal cells of the primate retina express GABA-like immunoreactivity [24] and contain the GABA-synthesizing enzyme glutamic acid decarboxylase (GAD 65) [77]. This suggests that, similar to horizontal cells in other vertebrates, they release GABA as their transmitter [85]. Although there are some reports of conventional vesicular synapses made by horizontal cells [39], most contacts made by horizontal cells lack morphological specializations expected in conventional synapses [85]. Indeed, it is known that GABA release from horizontal cells is Ca^{2+}-independent, non-vesicular, and carrier mediated [1, 68, 69]. GABA release from horizontal cells was remarkably similar to the properties of the GABA transporter GAT-1 stably expressed in a cell line [54]. Unfortunately, GAT-1 does not seem to be expressed in horizontal cells [34].

The present anatomical study adds some important details to the mechanism of GABA release by horizontal cells. In agreement with other mammalian retinae [14], we found that monkey horizontal cells express the vesicular GABA transporter VGAT (Fig. 8). This transporter showed a low level of expression all along the horizontal cell membrane, but was greatly enriched at the photoreceptor synaptic terminals. The lateral elements in cone pedicles and rod spherules and hot spots at the level of the desmosome-like junctions were most intensely labeled. Only electron microscopy together with postembedding immunolabeling can resolve the issue regarding the precise ultrastructural localization of VGAT to the plasma membrane or to vesicles inside the dendritic/axonal terminals. However, if we assume that VGAT plays some role in the release of GABA by horizontal cells, the specific accumulation of VGAT close to the dendritic and axonal terminals suggests a *local* release of GABA upon depolarization of the horizontal cells. The release of GABA from horizontal cells could also be cone-specific: since the glutamate receptors are aggregated underneath every cone pedicle, it is possible that they cause a small, local membrane depolarization that triggers the release of GABA at that particular cone. Only larger depolarizations would cause a depolarization of the whole cell and release of GABA at all cone pedicles contacted. Such local and thus cone-specific action of horizontal cells in the monkey retina is not mere speculation, because it has recently been shown by intracellular recordings that cone-specific sensitivity regulation occurs before summation of cone signals by horizontal cells [38].

The Localization of GABA Receptors

The pattern of aggregation of VGAT (Fig. 8G) would predict that GABA is released from the horizontal cells at their invaginating contacts and close to the desmosome-like junctions. GABA receptors on bipolar cell dendrites form a layer in between these two putative release sites (Fig. 8), suggesting that GABA could diffuse from either side to activate these receptors. Hence, there appears to be a spatial correspondence between the release of GABA and the aggregation of the GABA receptors at the bipolar cell dendrites.

Such feedforward action of GABA released from horizontal cells and acting on the bipolar cell dendrites faces a problem. In the case of OFF bipolar cells, light-evoked release of GABA from horizontal cells would cause "surround inhibition," which is the correct physiological signal. However, in ON bipolar cells, this would cause "surround excitation," which is the incorrect physiological signal. Vardi et al. [79] have recently immunostained retinae for the K-Cl cotransporter (KCC2) that normally extrudes chloride from the cytoplasm and for the Na-K-Cl cotransporter (NKCC) that normally accumulates chloride. KCC2 was expressed in OFF bipolar cell dendrites, and in OFF and ON bipolar cell axons. NKCC was expressed in ON bipolar cell dendrites. This would predict that E_{CL} is more positive than the resting potential in ON bipolar cell dendrites and release of GABA from horizontal cells would depolarize ON bipolar cell dendrites, which would be the correct physiological signal. Recent physiological recordings from bipolar cells in the mouse retina have partially confirmed this model by showing that GABA depolarizes rod bipolar cell dendrites [63].

It has to be emphasized that not only do the bipolar cells underneath the cone pedicles express GABA receptors, but physiological experiments have also shown that cone pedicles themselves respond to GABA [35, 82]. Although the molecular composition of the GABA Rs expressed by cone pedicles is still unknown, the local release of GABA, proposed in the present study, would also serve as an appropriate source for such cone pedicle GABA Rs.

References

1. Attwell D, Barbour B, Szatkowski M (1993) Nonvesicular release of neurotransmitter. Neuron 11:401–407
2. Awatramani GB, Slaughter MM (2000) Origin of transient and sustained responses in ganglion cells of the retina. J Neurosci 20:7087–7095

3. Barbour B, Häusser M (1997) Intersynaptic diffusion of neurotransmitter. Trends Neurosci 20:377–384
4. Bech-Hansen NT, Naylor MJ, Maybaum TA, et al. (1998) Loss-of-function mutations in a calcium-channel α1-subunit gene in Xp11.23 cause incomplete X-linked congenital stationary night blindness. Nat Genet 19:264–267
5. Boycott BB, Hopkins JM (1993) Cone synapses of a flat diffuse cone bipolar cell in the primate retina. J Neurocytol 22:765–778
6. Boycott BB, Wässle H (1991) Morphological classification of bipolar cells of the primate retina. Eur J Neurosci 3:1069–1088
7. Brandstätter JH, Hack I (2001) Localization of glutamate receptors at a complex synapse. Cell Tissue Res 303:1–14
8. Brandstätter JH, Koulen P, Wässle H (1997) Selective synaptic distribution of kainate receptor subunits in the two plexiform layers of the rat retina. J Neurosci 17:9298–9307
9. Brandstätter JH, Koulen P, Wässle H (1998) Diversity of glutamate receptors in the mammalian retina. Vision Res 38:1385–1397
10. Brandstätter JH, Fletcher EL, Garner CC, et al. (1999) Differential expression of the presynaptic cytomatrix protein bassoon among ribbon synapses in the mammalian retina. Eur J Neurosci 11:3683–3693
11. Calkins DJ, Tsukamoto Y, Sterling P (1996) Foveal cones form basal as well as invaginating junctions with diffuse ON bipolar cells. Vision Res 36:3373–3381
12. Chan TL, Grünert U (1998) Horizontal cell connections with short wavelength-sensitive cones in the retina: a comparison between new world and old world primates. J Comp Neurol 393:196–209
13. Chun M-H, Grünert U, Martin PR, et al. (1996) The synaptic complex of cones in the fovea and in the periphery of the macaque monkey retina. Vision Res 36:3383–3395
14. Cueva JG, Haverkamp S, Reimer RJ, et al. (2002) Vesicular γ-aminobutyric acid transporter expression in amacrine and horizontal cells. J Comp Neurol 445:227–237
15. Dacey DM, Lee BB, Stafford DK, et al. (1996) Horizontal cells of the primate retina: cone specificity without spectral opponency. Science 271:656–659
16. Derouiche A, Rauen T (1995) Coincidence of L-glutamate/L-aspartate transporter (GLAST) and glutamine synthetase (GS) immunoreactions in retinal glia: evidence for coupling of GLAST and GS in transmitter clearance. J Neurosci Res 42:131–143
17. DeVries SH (2000) Bipolar cells use kainate and AMPA receptors to filter visual information into separate channels. Neuron 28:847–856
18. Dingledine R, Borgs K, Bowie D, et al. (1999) The glutamate receptor ion channels. Pharmacol Rev 51:7–61
19. Dowling JE, Boycott BB (1966) Organization of the primate retina: electron microscopy. Proc R Soc Lond B Biol Sci 166:80–111
20. Enz R, Brandstätter JH, Wässle H, et al. (1996) Immunocytochemical localization of the GABA_C receptor ρ subunits in the mammalian retina. J Neurosci 16:4479–4490
21. Garner CC, Nash J, Huganir RL (2000) PDZ domains in synapse assembly and signalling. Trends Cell Biol 10:274–280
22. Gersdorff von H (2001) Synaptic ribbons: versatile signal transducers. Neuron 29:7–10
23. Greferath U, Grünert U, Müller F, et al. (1994) Localization of GABA_A receptors in the rabbit retina. Cell Tissue Res 276:295–307
24. Grünert U, Wässle H (1990) GABA-like immunoreactivity in the macaque monkey retina: a light and electron microscopic study. J Comp Neurol 297:509–524
25. Gumbiner BM (1996) Cell adhesion: the molecular basis of tissue architecture and morphogenesis. Cell 84:345–357
26. Hack I, Frech M, Dick O, et al. (2001) Heterogeneous distribution of AMPA glutamate receptor subunits at the photoreceptor synapses of rodent retina. Eur J Neurosci 13:15–24
27. Haverkamp S, Grünert U, Wässle H (2000) The cone pedicle, a complex synapse in the retina. Neuron 27:85–95

28. Haverkamp S, Grünert U, Wässle H (2001a) The synaptic architecture of AMPA receptors at the cone pedicle of the primate retina. J Neurosci 21:2488–2500

29. Haverkamp S, Grünert U, Wässle H (2001b) Localization of kainate receptors at the cone pedicles of the primate retina. J Comp Neurol 436:471–486

30. Hollmann M, Heinemann S (1994) Cloned glutamate receptors. Annu Rev Neurosci 17:31–108

31. Hopkins JM, Boycott BB (1997) The cone synapses of cone bipolar cells of primate retina. J Neurocytol 26:313–325

32. Honda S, Yamamoto M, Saito N (1995) Immunocytochemical localization of three subtypes of GABA transporter in rat retina. Brain Res Mol Brain Res 33:319–325

33. Hu M, Bruun A, Ehinger B (1999) Expression of GABA transporter subtypes (GAT1, GAT3) in the adult rabbit retina. Acta Ophthalmol Scand 77:255–260

34. Johnson J, Chen TK, Rickman DE, et al. (1996) Multiple γ-aminobutyric acid plasma membrane transporters (GAT-1, GAT-2, GAT-3) in the rat retina. J Comp Neurol 375:212–224

35. Kaneko A, Tachibana M (1986) Effects of gamma-aminobutyric-acid on isolated cone photoreceptors of the turtle retina. J Physiol (Lond) 373:443–461

36. Kennedy MB (1997) The postsynaptic density at glutamatergic synapses. Trends Neurosci 20:264–268

37. Koulen P, Fletcher EL, Craven SE, et al. (1998) Immunocytochemical localization of the postsynaptic density protein PSD-95 in the mammalian retina. J Neurosci 18:10136–10149

38. Lee BB, Dacey DM, Smith VC, et al. (1999) Horizontal cells reveal cone type-specific adaptation in primate retina. Proc Natl Acad Sci USA 96:14611–14616

39. Linberg KA, Fisher SK (1988) Ultrastructural evidence that horizontal cell axon terminals are presynaptic in the human retina. J Comp Neurol 268:281–297

40. McIntire SL, Reimer RJ, Schuske K, et al. (1997) Identification and characterization of the vesicular GABA transporter. Nature 389:870–876

41. Missotten L (1965) The ultrastructure of the human retina. Editions Arscia S.A., Brussels

42. Miyake Y, Yagasaki K, Horiguchi M, et al. (1986) Congenital stationary night blindness with negative electroretinogram: a new classification. Arch Ophthalmol 104:1013–1020

43. Morgans CW (2000a) Presynaptic proteins of ribbon synapses in the retina. Microsc Res Tech 50:141–150

44. Morgans CW (2000b) Neurotransmitter release at ribbon synapses in the retina. Immunol Cell Biol 78:442–446

45. Morgans CW (2001) Localization of the α1F calcium channel subunit in the rat retina. Invest Ophthalmol Vis Sci 42:2414–2418

46. Morigiwa K, Vardi N (1999) Differential expression of ionotropic glutamate receptor subunits in the outer retina. J Comp Neurol 405:173–184

47. Muresan V, Lyass A, Schnapp BJ (1999) The kinesin motor KIF3A is a component of the presynaptic ribbon in vertebrate photoreceptors. J Neurosci 19:1027–1037

48. Nachman-Clewner M, St. Jules R, Townes-Anderson E (1999) L-type calcium channels in the photoreceptor ribbon synapse: localization and role in plasticity. J Comp Neurol 415:1–16

49. Nomura A, Shigemoto R, Nakamura Y, et al. (1994) Developmentally regulated postsynaptic localization of a metabotropic glutamate receptor in rat rod bipolar cells. Cell 77:361–369

50. Ozawa S, Haruyuki K, Tsuzuki K (1998) Glutamate receptors in the mammalian central nervous system. Prog Neurobiol 54:581–618

51. Petralia RS, Wang YX, Wenthold RJ (2002) NMDA receptors and PSD-95 are found in attachment plaques in cerebellar granular layer glomeruli. Eur J Neurosci 15:583–587

52. Pow DV, Barnett NL (1999) Changing pattern of spatial buffering of glutamate in developing rat retinae are mediated by the Müller cell glutamate transporter GLAST. Cell Tissue Res 297:57–66
53. Qin P, Pourcho RG (1999) Localization of AMPA-selective glutamate receptor subunits in the cat retina: a light- and electron microscopic study. Vis Neurosci 16:169–177
54. Rakhilin SV, Schwartz EA (1994) A GABA transporter operates asymmetrically and with variable stoichiometry. Neuron 13:949–960
55. Rao-Mirotznik R, Harkins AB, Buchsbaum G, et al. (1995) Mammalian rod terminal: architecture of a binary synapse. Neuron 14:561–569
56. Rauen T, Taylor WR, Kuhlbrodt K, et al. (1998) High-affinity glutamate transporters in the rat retina: a major role of the glial glutamate transporter GLAST-1 in transmitter clearance. Cell Tissue Res 291:19–31
57. Raviola E, Gilula NB (1975) Intramembrane organization of specialized contacts in the outer plexiform layer of the retina. J Cell Biol 65:192–222
58. Rieke F, Schwartz EA (1994) A cGMP-gated current can control exocytosis at cone synapses. Neuron 13:863–873
59. Röhrenbeck J, Wässle H, Boycott BB (1989) Horizontal cells in the monkey retina: immunocytochemical staining with antibodies against calcium binding proteins. Eur J Neurosci 1:407–420
60. Sagné C, Mestikawy SE, Isambert M-F, et al. (1997) Cloning of a functional vesicular GABA and glycine transporter by screening of genome databases. FEBS Lett 417:177–183
61. Sandmann D, Boycott BB, Peichl L (1996) The horizontal cells of artiodactyl retinae: a comparison with Cajal's descriptions. Vis Neurosci 13:735–746
62. Sarantis M, Mobbs P (1992) The spatial relationship between Müller cell processes and the photoreceptor output synapse. Brain Res 584:299–304
63. Satoh H, Kaneda M, Kaneko A (2001) Intracellular chloride concentration is higher in rod bipolar cells than in cone bipolar cells of the mouse retina. Neurosci Lett 310:161–164
64. Schmitz Y, Witkovsky P (1997) Dependence of photoreceptor glutamate release on a dihydropyridine-sensitive calcium channel. Neuroscience 78:1209–1216
65. Schmitz F, Königstorfer A, Südhof TC (2000) RIBEYE, a component of synaptic ribbons: a protein's journey through evolution provides insight into synaptic ribbon function. Neuron 28:857–872
66. Schneeweis DM, Schnapf JL (1995) Photovoltage of rods and cones in the macaque retina. Science 268:1053–1056
67. Schneeweis DM, Schnapf JL (1999) The photovoltage of macaque cone photoreceptors: adaptation, noise, and kinetics. J Neurosci 19:1203–1216
68. Schwartz EA (1993) Depolarization without calcium can release γ-aminobutyric acid from a retinal neuron. Science 238:350–355
69. Schwartz EA (1999) A transporter mediates the release of GABA from horizontal cells. In: Toyoda J-I, et al. (eds) The retinal basis of vision. Elsevier, New York, pp 93–101
70. Strom TM, Nyakatura G, Apfelstedt-Sylla E, et al. (1998) An L-type calcium-channel gene mutated in incomplete X-linked congenital stationary night blindness. Nat Genet 19:260–263
71. Taylor WR, Morgans CW (1998) Localization and properties of voltage-gated calcium channels in cone photoreceptors of *Tupaia belangeri*. Vis Neurosci 15:541–552
72. Thoreson WB, Witkovsky P (1999) Glutamate receptors and circuits in the vertebrate retina. Prog Retin Eye Res 18:765–810
73. tom Dieck S, Sanmarti-Vila L, Langnaese K, et al. (1998) Bassoon, a novel zinc-finger CAG/glutamine-repeat protein selectively localized at the active zone of presynaptic nerve terminals. J Cell Biol 142:499–509

74. Vandenbranden CAV, Verweij J, Kamermans M, et al. (1996) Clearance of neurotransmitter from the cone synaptic cleft in goldfish retina. Vision Res 36:3859–3874
75. Vardi N, Sterling P (1994) Subcellular localization of $GABA_A$ receptor on bipolar cells in macaque and human retina. Vision Res 34:1235–1246
76. Vardi N, Masarachia P, Sterling P (1992) Immunoreactivity to $GABA_A$ receptor in the outer plexiform layer of the cat retina. J Comp Neurol 320:394–397
77. Vardi N, Kaufman, DL, Sterling P (1994) Horizontal cells in cat and monkey retina express different isoforms of glutamic acid decarboxylase. Vis Neurosci 11:135–142
78. Vardi N, Morigiwa K, Wang T-L, et al. (1998) Neurochemistry of the mammalian cone "synaptic complex." Vision Res 38:1359–1369
79. Vardi N, Zhang L-L, Payne JA, et al. (2000) Evidence that different cation chloride cotransporters in retinal neurons allow opposite responses to GABA. J Neurosci 20:7657–7663
80. Wässle H, Dacey DM, Haun T, et al. (2000) The mosaic of horizontal cells in the macaque monkey retina: with a comment on biplexiform ganglion cells. Vis Neurosci 17:591–608
81. Wilkinson MF, Barnes S (1996) The dihydropyridine-sensitive calcium channel subtype in cone photoreceptors. J Gen Physiol 107:621–630
82. Wu SM (1994) Synaptic transmission in the outer retina. Annu Rev Physiol 56:141–168
83. Wu SM, Gao F, Maple BR (2000) Functional architecture of synapses in the inner retina: segregation of bipolar cell axon terminals. J Neurosci 20:4462–4470
84. Yagi T, MacLeish P (1994) Ionic conductances of monkey solitary cone inner segments. J Neurophysiol 71:656–665
85. Yazulla S (1995) Neurotransmitter release from horizontal cells. In: Djamgoz, et al. (eds) Neurobiology and clinical aspects of the outer retina. Chapman and Hall, London, pp 249–271

Transmission at the Mammalian Cone Photoreceptor Basal Synapse

Steven H. DeVries

Key words. Cone photoreceptor, Glutamate receptor, Bipolar cell

Cone photoreceptor synaptic terminals have both an unusual structure and an unusual function. The unusual structure is called the basal contact. At these contacts, the membranes of a cone and a postsynaptic cell, usually an Off bipolar cell, come into close apposition, but the stigmata of synaptic transmission, presynaptic clusters of vesicles and active zones, are absent. Rather, the cone transmitter, glutamate, is thought to be released solely at synaptic ribbons located atop invaginations, and must diffuse 200–500 nm to the nearest basal contacts. The unusual function is that the cone synapse transmits voltage signals that are hyperpolarizing, graded, and frequently very small. How does the unusual structure of the basal contact relate to its function in transmission?

Transmission at the cone basal synapse was studied by recording simultaneously in voltage clamp from a cone and a postsynaptic Off bipolar cell in slices from the cone-dominant ground squirrel retina *(Spermophilus tridecemlineatus)*. By voltage clamping pre- and postsynaptic cells, it was possible to elicit and resolve quantal transmission events whose time course might be limited by diffusion. Unexpectedly, the different anatomical types of Off bipolar cells had differently shaped quantal events. Events in b2 bipolar cells were consistently large and fast (10–20 pA peak, 100–180 μs, 20%–80% rise time), whereas those in b3 and b7 bipolars were slower and fell into two groups (5–10 pA peak with 300–600 μs rise and 1–3 pA peak with 800–1800 μs rise). The differences in response kinetics did not appear to result from the properties of postsynaptic receptors, the cable properties of bipolar cell dendrites, or the existence of distinct pools of releasable vesicles. Rather, differences in quantal response shape were attributed to characteristic differences in diffusion distance: b2 bipolar cells contact cones closer to ribbon release sites, perhaps preferentially occupying the triad-associated position, while b3 and b7 cells preferentially make their contacts farther from release sites. The impact of quantal response shape on the noise of transmission at the cone synapse is currently under investigation.

Departments of Ophthalmology and Physiology, Northwestern University Medical School, 303 East Chicago Ave., Tarry 5-715, Chicago, IL 60611, USA

Impaired Transmission from Photoreceptors to Bipolar Cells: Mouse Models

NEAL S. PEACHEY

Key words. Mouse, Electroretinogram, Retina

In response to light, photoreceptors normally decrease the rate of glutamate release at their terminal. This change is detected by two classes of bipolar cells, which are distinguished by whether they depolarize (DBCs) or hyperpolarize (HBCs) in response to light. The electroretinogram (ERG) can be used to monitor this process as the ERG a-wave reflects photoreceptor activity while the b-wave is generated by bipolar cell responses. As a consequence, a selective reduction in the ERG b-wave is usually indicative of a defect in transmission between photoreceptors and bipolar cells. This situation is encountered in naturally occurring and genetically engineered mice, the study of which shed light on the development and maintenance of the photoreceptor-to-bipolar cell synaptic process. For example, null mutations in the genes encoding mGluR6 or $G\alpha_0$ eliminate the b-wave and confirm the role of these proteins in DBC signal transduction. A similar ERG pattern is seen in the *nob (no b-wave)* mouse, which is caused by a defect in *nyx*, encoding nyctalopin. Although the precise function of nyctalopin is unknown, the localization of nyctalopin transcripts to the outer nuclear layer indicates that this protein may play a critical role in the DBC response. Despite the lack of DBC function, ribbon synapses form normally in *nob* mice. Presynaptically, glutamate release is controlled by L-type calcium channels located on the photoreceptor terminal. The b-wave is selectively reduced in $CNS\beta_2$-null mice, lacking the β_2 subunit of the L-type calcium channel in the central nervous system. In addition, the normal distribution of the pore-forming α_{1F} subunit is markedly disturbed, indicating that the β_2 subunit is required to guide the formation of a functional channel. Such channels are likely to be required for ribbon synapse formation, as ribbon structures are rarely encountered in $CNS\beta_2$-null mice. As the *nob* and $CNS\beta_2$-null mice involve genes that underlie two forms of congenital stationary night blindness, study of these animal models will expand our understanding of these rare human disorders.

Cole Eye Institute, Cleveland Clinic Foundation, 9500 Euclid Ave., Cleveland, OH 44195, USA
Research Service, Cleveland VA Medical Center, Cleveland, OH, USA

Selective Dysfunction of On- and/or Off-Bipolar Cell in Human Diseases

Yozo Miyake, Mineo Kondo, Makoto Nakamura,
and Hiroko Terasaki

Key words. Bipolar cell dysfunction, ERG, Psychophysical function

In human, cone visual signals in the retina are processed through dual pathways, one involving ON-center bipolar cells (ON-pathway) and the other through OFF-center bipolar cells (the OFF-pathway), while rod visual signal is processed only through the ON-pathway. Little is known about how patients can see when the ON- and/or OFF-pathway is selectively defective. By comparing the monkey's electroretinogram (ERG) of a blocked ON- or OFF-pathway using glutamate neurotransmitter analogs, we have detected new clinical entities in human where ON- and/or OFF-pathways are blocked in both rod and cone visual pathways. We hypothesized that the complete type of congenital stationary night blindness (CSNB 1) can be a model disease, having a complete dysfunction of only the ON-pathway, while the incomplete type (CSNB 2) has an incomplete dysfunction of both the ON- and OFF-pathways in both rod and cone visual systems. The gene mutation was detected in nyctalopin in CSNB 1, and in L-type calcium channel $\alpha 1$ subunit in CSNB 2.

We analyzed 90 patients (49 patients with CSNB 1, 41 patients with CSNB 2) in terms of scotopic vision (dark adaptation) and photopic vision (visual acuity, contrast sensitivity, and color vision). The scotopic vision was found to be impaired completely and incompletely in CSNB 1 and CSNB 2, respectively. Photopic vision was better preserved than we expected even when the full-field photopic ERG showed a complete defect of the ON-visual pathway. Judging from the findings of focal macular cone ERG, we found that the macula of these patients is selectively spared from the generalized defect of the entire retina.

Department of Ophthalmology, Nagoya University School of Medicine, 65 Tsurumai-cho, Showa-ku, Nagoya 466-8550, Japan

Neural Circuit Contributing to the Formation of the Receptive Surround

Akimichi Kaneko and Hajime Hirasawa

Key words. Cone photoreceptor, Horizontal cell, Receptive field surround

Formation of the center-surround organization of the receptive field is one of the most important functions of the vertebrate retina, and many studies have been reported on its mechanism. Since the historical finding of [1] in turtle cones, it has been widely accepted that the feedback signal from the horizontal cell (HC) is responsible for the formation of the antagonistic surround response of cone photoreceptors. Accumulating evidence has suggested that the feedback from HC to cones is mediated by γ-aminobutyric acid (GABA); HCs are GABAergic, they release GABA upon depolarization, and GABA receptors are concentrated at the cone terminal. On the other hand, there are number of reports criticizing the GABA hypothesis of HC–cone feedback.

To find an answer to this controversial question, we carried out a whole-cell recording from cones in a slice preparation of the newt retina. The pipette solution contained 20 mM 1,2-bis(2-aminophenoxy)ethane-N,N,N1,N1-tetraacetic acid (BAPTA) to suppress Ca^{2+}-induced currents in cones. Illumination of the cone with a 30-μm light spot evoked hyperpolarization. Superposition of diffuse illumination over the spot evoked depolarization. These light responses remained unchanged in the presence of 100 μM picrotoxin, but were abolished by 20 μM kainate (HC depolarized) or by 20 μM 6-cyano-7-nitroquinoxaline (CNQX) (HC hyperpolarized). These observations suggest that HC is actively involved in the surround response of cones, but without using GABA.

During the surround effect, the Ca^{2+} current (I_{Ca}) in cones was enhanced, and the enhancement was voltage dependent; there was strong enhancement in the voltage range between −20 mV (the voltage at which I_{Ca} was maximum) and −40 mV, but no enhancement at voltages more positive than +10 mV. Such enhancement is seen when the extracellular solution is made more basic. Substitution of the superfusate containing bicarbonate buffer by a stronger pH buffer (the control solution supplemented with 5 mM hydroxyethylpiperazine ethanesultonic acid of identical pH) abolished the

Department of Physiology, School of Medicine, Keio University, 35 Shinanomachi, Shinjukn-ku, Tokyo 160-8582, Japan

surround effect reversibly. In the presence of kainate the I_{Ca} of cones was suppressed, while in the presence of CNQX the I_{Ca} of cones was enhanced. From these observations, we hereby propose a new hypothesis that I_{Ca} of cones, which regulates the transmitter release from cones, is affected by the pH of the extracellular solution in the invaginating synaptic cleft, which is controlled by the membrane voltage of HC. The molecular mechanism that controls the proton movement in relation to the HC membrane voltage is still open and needs further studies.

Acknowledgments. This work was supported by Grant-in-Aid for Scientific Research from Japan Society for the Promotion of Science, Japan.

Reference

1. Baylor DA, Fuortes MG, O'Bryan PM (1971) Receptive fields of cones in the retina of the turtle. J Physiol 214:265–294

GABAergic Modulation of Ephaptic Feedback in the Outer Retina

Maarten Kamermans, Iris Fahrenfort, Trijntje Sjoerdsma, and Jan Klooster

Key words. Cones, Horizontal cells, Feedback, GABA

In goldfish, both cones and horizontal cells (HCs) have ionotropic γ-aminobutyric acid (GABA) receptors [1, 2], and HCs release GABA using a GABA transporter [3, 4]. It has been proposed that HCs feed back to cones via this GABAergic system. However, full-field light stimulation does not lead to a closure of a GABA-gated conductance in the cones. Instead, negative feedback from HCs to cones is mediated via an ephaptic mechanism involving hemichannels [5]. This leaves the role of the GABAergic system, present in the outer retina, unexplained.

Here we show that the GABAergic system modulates the amplitude of the feedback-mediated responses in both cones and HCs. This modulation depends on the GABA receptors in the cones, since changing the chloride equilibrium potential in cones to $-20\,mV$ abolishes the modulatory effect of GABA. Estimates of the time constant of the modulatory effect of GABA indicate that it is larger than $3\,s$, showing that this pathway is much slower than the ephaptic feedback pathway from HCs to cones. Immunolocalization of GAT1, GAT2, and GAT3 revealed that the majority of the transporters on HC are outside the cone/HC synapse, whereas part of the GABA receptors are inside the synaptic terminal.

To account for all of these results, we propose the following mechanism. Current flows into the hemichannels on the horizontal cell dendrites. This current induces a voltage drop along the intersynaptic space in the synaptic terminal, making the extracellular potential near the Ca channels slightly negative. This forms the basis for the ephaptic negative feedback mechanism. Since part of the GABA-gated channels of the cone are in the cone/HC synapse, the GABA-gated current flowing out of the cone can provide part of the hemichannel current, and thereby decrease the amount of current flowing through the intersynaptic space from outside the synaptic terminal. This GABAergic interaction with the hemichannel-mediated current leads to a decrease in voltage drop in the intersynaptic space, leading to a decrease of feedback. The relative large distance between the GABA transporters and the GABA receptors suggests

Research Unit Retinal Signal Processing, The Netherlands Ophthalmic Research Institute, Amsterdam, The Netherlands

that HCs release GABA in a large extrasynaptic compartment. This organization might account for the slow time constant of the GABAergic modulation.

Since the GABA release is high in the dark-adapted retina, the function of this GABAergic system might be to down-regulate feedback in the dark-adapted condition.

References

1. Kaneko A, Tachibana M (1986) Effects of gamma-aminobutyric acid on isolated cone photoreceptors of the turtle retina. J Physiol 373:443–461
2. Verweij J, Kamermans M, Negishi K, et al. (1998) GABA sensitivity of spectrally classified horizontal cells in goldfish retina. Vis Neurosci 15:77–86
3. Yazulla S, Kleinschmidt J (1982) Dopamine blocks carrier-mediated release of GABA from retinal horizontal cells. Brain Res 233:211–215
4. Schwartz EA (1987) Depolarization without calcium can release gamma-aminobutyric acid from a retinal neuron. Science 238:350–355
5. Kamermans M, Fahrenfort I, Schultz K, et al. (2001) Hemichannel-mediated inhibition in the outer retina. Science 292:1178–1180

SN1 Catalyzes Transport of Glutamine Across Müller Cells in Retina

Kyoh-ichi Takahashi[1,2], David Krizaj[2], and David R. Copenhagen[2]

Key words. Müller cell, Glutamine transport, Glutamate/glutamine cycle

High-affinity glutamate (Glu) transporters localized on Müller cell membranes transport synaptically released Glu into these cells from the extracellular side. Müller cells, similar to most glial cells in the nervous system, generate glutamine (Gln) from Glu using the enzyme glutamine synthetase. Although it has been postulated that Gln is transported from these Müller cells to replenish synaptic Glu, little is known about the identity or properties of the Gln transporter. We investigated the Gln transport mechanism in Müller cells dissociated from retinae of tiger salamander (*Ambystoma tigrinum*) using three different approaches: intracellular pH (pH_i) measurements, whole-cell voltage-clamp, and immunocytochemistry.

Gln changed pH_i in Müller cells, alkalinizing them at millimolar concentrations. Gln-evoked alkalinizations were suppressed when $[Na^+]_o$ was replaced by $[NMDG^+]_o$ but not by $[Li^+]_o$. These findings suggest a Na^+-dependent mechanism associated with Gln/H^+ exchange. Three types of Gln transporter played no role in the Gln-induced alkalinizations. The alkalinizations were not blocked by threonine (4 mM), a blocker of System A, or by a-methylaminoisobutyric acid (3 mM), a blocker of System ASC, or by 2-amino-2-norborane-carboxylic acid (3 mM), a blocker of System L. These pH_i measurements strongly suggest that the recently cloned System N family of Gln transporters (SN1; [1]) mediates Gln transport across retinal Müller cells. Further evidence was provided from immunocytochemical analysis. Antibodies to SN1 stained Müller cells.

Voltage-clamp studies of Gln-induced currents further implicated SN1. Recent studies of SN1 expressed in *Xenopus* oocytes showed that an uncoupled current is activated by SN1-mediated Gln and asparagine (Asn) transport. We found that, in Müller cells, Gln and, to a greater extent, Asn evoked inward currents that were larger, at more hyperpolarized membrane potentials. The I–V relation is roughly linear and no currents were evoked at around +40 mV. These properties are consistent with a

[1] Division of Biological Sciences, Faculty of Human Environmental Studies, Hiroshima Shudo University, 1-1-1 Otsuka-higashi, Asaminami-ku, Hiroshima 731-3195, Japan
[2] Departments of Ophthalmology and Physiology, UCSF School of Medicine, San Francisco, CA 94143-0730, USA

transporter-mediated current selective for Na^+ and/or H^+. Similar to the characteristics of SN1 found in the oocyte expression system, high extracellular pH (pH_o) enhanced Gln and Asn currents and low pH_o suppressed these currents in the Müller cells. These results are consistent with an uncoupled current activated by glutamine transport via SN1.

Based on results from three different types of experiment, we conclude that Gln transport in retinal Müller cells is mediated by an SN1-like transporter. It is likely that this transporter plays a role in the Glu/Gln recycling pathway in which Gln is pumped back into the glutamatergic neurons where it becomes a substrate for Glu synthesis.

Reference

1. Chaudhry FA, Reimer RJ, Krizaj D, et al. (1999) Molecular analysis of System N suggests novel physiological roles in nitrogen metabolism and synaptic transmission, Cell 99:769–780

Some Functions of Amacrine Cells

Richard H. Masland

Key words. Amacrine cell, Ganglion cell, Receptive field

In contrast to traditional schemas, amacrine cells make up a major population of neurons in the retina, both quantitatively and in the wide variety of their types. They represent the major input to retinal ganglion cells, occupying a fraction ranging from 70% for alpha cells in the cat and parasol cells in the monkey, to 50% for midget ganglion cells in the monkey fovea. In addition, amacrine cells make major inhibitory synapses on the axon terminals of the bipolar cells, thus controlling their output to the ganglion cells. They thus are major players in shaping the flow of information through the retina. In contrast, horizontal cells come in only two types and seem to have a single, unified function. Accordingly, amacrine cells outnumber horizontal cells by numbers that range from 4:1 to 10:1.

One estimate of the number of amacrine cell types is that 29 distinct cell types exist. This is clearly not definitive and a few rare additional cells are sure to be discovered; but it does eliminate the possibility that any large single class has been missed. Certain generalizations emerge from amacrine cell structure. There are many narrow-field amacrine cells (field diameter less than 175 μm) and these cells create a local microstructure within the receptive fields of most ganglion cells. Conversely, all retinas studied thus far contain a large array of wide-field cells, sometimes with dendrites that run many millimeters across the retina. These are presumably responsible for the periphery or shift effects, and for long-range oscillations, but their function remains to be studied in more detail. Finally, a group of complex medium-field cells exists; their field diameters cluster around 175 μm and they usually span several layers of the inner plexiform layer.

What candidate functions might be imagined for this array of amacrine cells? One is correlated firing among ganglion cells, which may convey information about features of the visual scene. Some of the wide-field cells are modulators, which control the responsiveness of many retinal elements, particularly during light and dark adaptation. Others participate in highly specific functions such as direction selectivity. A

Howard Hughes Medical Institute, Massachusetts General Hospital, Harvard Medical School, Boston, MA 02114, USA

fourth class contribute to contrast gain control, in its several different forms. It is now clear that contrast gain control is not a unitary function; there is much evidence for multiple different forms of contrast adaptation tuned to different features of the visual stimulus. Finally, there is evidence for more complex and subtle long-range effects in the retina that are revealed simply by the periphery and shift effects. One such function may include anticipatory responses to moving stimuli; another is a saccadic shutting down of the retina's output during eye movements.

Synaptic Mechanisms Underlying Direction Selectivity in Rabbit Retinal Ganglion Cells

W. Rowland Taylor[1] and David I. Vaney[2]

Key words. Direction selectivity, Ganglion cells, Synaptic conductance

In the vertebrate visual system, direction-selective signals are first encountered within the direction-selective ganglion cells (DSGCs) in the retina. The On-Off DSGCs have been most intensively studied, and a number of models have been advanced to account for the physiological properties of these cells. In order to account for the directional asymmetry of the physiological responses, all models incorporate some type of spatial asymmetry in the arrangement of the synaptic circuitry. However, there is no general agreement as to the synaptic locus of this critical synaptic asymmetry. Some have suggested that direction selectivity is generated by synaptic interactions that occur prior to the DSGC, and that the excitatory or inhibitory inputs to the cell are already direction selective. Others have suggested that the inputs to the DSGCs are directionally isotropic, and that there is an asymmetry in the interaction of the excitation and inhibition within the dendrites of the DSGCs.

We have measured synaptic conductances in the DSGCs from rabbit retina in an effort to determine whether the synaptic inputs to the DSGCs are already directional. Responses were elicited by stimuli moving to and fro along the preferred-null axis of the cell at a range of membrane potentials between −100 and +20 mV. Synaptic conductance was estimated from the slope of the current-voltage relations of the net light-evoked synaptic currents. Conductance was estimated as a function of time during the synaptic responses for both the preferred and null direction motion. The resulting conductance records were then separated into components due to excitation and inhibition.

The study provides evidence for three types of synaptic asymmetry that contribute to generating direction-selective responses. Inhibition impinging upon the DSGCs is stronger in the null than in the preferred direction. Conversely, excitation is stronger in the preferred direction than in the null direction. Thus, both excitatory and inhibitory inputs to the cells appear to be directional, indicating the existence of

[1] Neurological Sciences Institute, Oregon Health and Science University, Portland, OR 97239-3098, USA
[2] Vision, Touch and Hearing Research Centre, School of Biomedical Science, University of Queensland, Brisbane, Queensland, Australia

synaptic asymmetries within the circuitry prior to the DSGC. A third synaptic asymmetry appears to be postsynaptic and manifests as a delay in the inhibition in the preferred direction, with the result that excitation and inhibition are coincident in the null direction, but temporally disparate in the preferred direction. Surprisingly, this third postsynaptic asymmetry was evident only for the off-responses and was absent for the on-responses. Therefore, although the on- and off-dendritic strata are functionally equivalent, the underlying synaptic mechanisms are diverse.

Processing and Integration Mechanisms of Visual Transmission in the Retinal Network

Shigetada Nakanishi

Key words. Retinal network, ON/OFF segregation, Directional selectivity

Visual signals are transmitted in the retinal network through the principal pathway consisting of photoreceptors, bipolar cells, and ganglion cells. The key processing of visual transmission is segregation of light and dark signals into ON and OFF responses at the level of bipolar cells. This segregation is interesting because glutamate transmitter is used as a common transmitter but generates opposite ON and OFF responses, depending on bipolar cell types. We identified and characterized mGluR6 metabotropic glutamate receptor subtype in the retina. mGluR6 is selectively and exclusively expressed at postsynaptic sites of ON bipolar cells, and mGluR6 deficiency by gene targeting causes a loss of ON responses without affecting OFF responses. Two distinct types of glutamate receptors, metabotropic and ionotropic glutamate receptors, are thus effectively used for segregating light and dark signals into ON and OFF responses in the retinal network.

In the retinal network, a variety of amacrine cells exists and differently modulates synaptic transmission in the principal retinal pathway. Starburst cell is one of the amacrine cells and forms an asymmetric connection with bipolar cells and ON-OFF ganglion cells. ON-OFF ganglion cells show direction-selective responses to a moving light stimulus. To explore the role of starburst cells in directional selectivity, we specifically ablated starburst cells from the retinal network, using immunotoxin-mediated cell targeting techniques. Elimination of starburst cells not only abolishes direction-selective responses of ON-OFF ganglion cells but also causes a marked deficit in driving eye movements that respond to a moving stimulus. Pharmacological characterization of starburst cell-eliminated retina indicated that γ-aminobutyric acid released from starburst cells is critical in generating directional selectivity, most likely by inhibiting glutamate receptors in ON-OEF ganglion cells. Visual signals are thus transmitted in a concerted and spatiotemporal manner and this integrative synaptic transmission plays an important role in information processing of visual transmission.

Department of Biological Sciences, Kyoto University, Faculty of Medicine, Yoshida, Sakyo-ku, Kyoto 606-8501, Japan

Narrow-Field Amacrine Cells Mediate OFF Signals Between B2 Bipolar and Ganglion Cells in the Mouse Retina

Yoshihiko Tsukamoto

Key words. Mouse retina, Amacrine cell, Microcircuitry

A rod-cone mixed input type of OFF bipolar cell (B2) comprises the third rod signal pathway in the mouse retina [1]. The B2 cell receives cone input at basal contacts and also rod input at symmetrical contacts that express ionotropic glutamate receptors [2]. Only one fifth of all rod terminals have direct contacts with B2 cell dendrites, but such rod members gather signals from neighboring rods through rod–rod gap Junctions. This third pathway explains the rod responses that persist in OFF ganglion cells after blockage of the other pathways [3].

Examination of a series of 366 radial sections of the mouse retina has revealed small and large types of OFF ganglion cells that receive the B2 input at ribbon synapses. In addition, a type of narrow-field amacrine cell (like Kolb's type A4, 1981) relays the B2 signal to the small OFF ganglion cell. The dendrites of this A4-like amacrine cell are nearly as narrow as those of the AII amacrine cell but monostratified within the outer strata of the inner plexiform layer (IPL). Also, the other type of narrow-field amacrine cell (like Kolb's type A3) relays the B2 signal to the large OFF ganglion cell. The dendrites of this A3-like amacrine cell are about twice wider in diameter than those of the A4-like cell. The pharmacological identity, excitatory or inhibitory, at its output synapses remains to be examined.

With increasing photon density, the B2 cell is thought to pool OFF signals from many rods, thus reducing noise by averaging [4]. Ribbon synapses at bipolar terminals may transfer the bipolar's graded potential to the ganglion cell. One reason for narrow-field amacrine cells to mediate additional connections to ganglion cells could be to emphasize temporal changes in the bipolar potential for responding to the front edge of a dark shadow entering the visual field.

References

1. Tsukamoto Y, Morigiwa K, Ueda M, et al. (2001) Microcircuits for night vision in mouse retina. J Neurosci 21:8616–8623

Department of Biology, Hyogo College of Medicine, 1-1 Mukogawa Nishinomiya, Hyogo 663-8501, Japan

2. Hack I, Peichl L, Brandstätter JH (1999) An alternative pathway for rod signals in the rodent retina: Rod photoreceptors, cone bipolar cells, and the localization of glutamate receptors. Proc Natl Acad Sci USA 96:14130–14135
3. Soucy E, Nirenberg S, Nathans J, et al. (1998) A novel signaling pathway from rod photoreceptors to ganglion cells in mammalian retina. Neuron 21:481–493
4. Field GD, Rieke F (2002) Mechanisms regulating variability of the single photon responses of mammalian rod photoreceptors. Neuron 35:733–747

Functional Roles of Action Potentials and Na Currents in Amacrine Cells

Shu-Ichi Watanabe[1], Amane Koizumi[2], Yoshitake Yamada[2], and Akimichi Kaneko[2]

Key words. Amacrine cell, Action potential, Na current, Inhibition, GABA

Introduction

Amacrine cells are retinal interneurons that play important roles in information processing in the inner plexiform layer. It is known that the major population of amacrine cells are γ-aminobutyric acid (GABA)-ergic or glycinergic. However, of over 20 morphological subtypes [1], functional roles are known in only two subtypes: glycinergic A2 amacrine cells and cholinergic starburst amacrine cells. GABAergic cells are thought to be inhibitory and their major roles are thought to send inhibitory feedback to bipolar cells, mutual inhibition to neighboring amacrine cells, and feed-forward inhibition to ganglion cells. Most amacrine cells lack an axon and their dendrites function not only as the input site but also as the output site. Therefore, the strength of inhibition is expected to depend on the magnitude of depolarization and the length of its propagation within dendritic processes. Here, we summarize our recent works on the functional role of the action potential and the sustained depolarization induced by transient and persistent Na currents in the information processing of amacrine cells.

The Role of the Action Potential in Amacrine Cells

Light responses of amacrine cells consist of graded polarization (either depolarization or hyperpolarization) and action potentials [2, 3]. It has been reported that the surround inhibition in ganglion cells mediated by large field amacrine cells is blocked by tetrodotoxin (TTX) [4, 5]. These results suggest that the action potential in amacrine cell dendrites is important to the surround inhibition. We have reported that

[1] Department of Physiology, Saitama Medical School, 38 Morohongo, Moroyamamachi, Saitama 350-0495, Japan
[2] Department of Physiology, Keio University School of Medicine, 35 Shinanomachi, Shinjuku-ku, Tokyo 160-8582, Japan

most amacrine cells receive GABAergic inputs from neighboring amacrine cells [6]. Spontaneous inhibitory postsynaptic currents (IPSC) mediated by GABA$_A$ receptor were observed by whole-cell recording in most amacrine cells of the goldfish retinal slice. Most IPSCs were miniature IPSCs (mIPSC, about 16 pA in amplitude at −80 mV), but large size IPSCs exceeding 100 pA were also recorded. TTX blocked only the large IPSC but not the mIPSC. These results suggest that the action potential generated in neighboring amacrine cells causes the large IPSC by a large amount of GABA release. Thus, the action potential of amacrine cells plays an important role in the information processing by sending a strong inhibition to distant postsynaptic cells.

Na Currents of Amacrine Cells

In amacrine cells that generate action potentials, a fast transient Na current (fast I_{Na}) was observed [7, 8]. We found a persistent (noninactivating) Na current (I_{NaP}) in amacrine cells of the goldfish retinal slice more frequently than the fast I_{Na} [8]. To investigate I_{NaP} further, we analyzed the current using cultured rat GABAergic amacrine cells [9]. Both the fast I_{Na} and I_{NaP} were recorded. It turned out that the activation voltage of I_{NaP} was at a more hyperpolarized level (between −60 and −50 mV) than that of the fast I_{Na} (between −40 and −30 mV). As the resting membrane potential of amacrine cells were about −65 mV, depolarization by 5 mV or more by an excitatory input is expected to activate I_{NaP}. In current clamp experiments, injection of a weak depolarizing current through the recording electrode caused depolarization of a few millivolts. The amplitude of depolarization was larger than that expected from the passive property of the membrane. Also, the evoked depolarization lasted after the current injection was switched off. The enhancements in amplitude and duration were blocked by TTX. Therefore, we concluded that I_{NaP} enhanced the depolarization. As amacrine cells possess Ca currents (I_{Ca}, activation voltage about −60 mV) [9, 10], application of I_{Ca} blockers also suppressed the enhancement [9].

Injection of a strong depolarizing current evoked action potentials superimposed on a large depolarization. Consistent results were observed in amacrine cells of the isolated retina of the carp. Light responses of amacrine cells consisted of graded depolarization and action potentials. TTX suppressed not only the action potentials but also the graded depolarization [8]. Thus, in these amacrine cells, I_{NaP}, together with I_{Ca}, enhanced the depolarization and facilitated the activation of fast I_{Na}, leading to the generation of action potentials.

Control of Action Potential Propagation in Amacrine Cell Dendrites

It has been suggested that the action potentials in amacrine cells propagate in the dendrites [11]. To confirm this notion more directly, we performed double whole-cell patch clamping in cultured GABAergic amacrine cells. Using the action potential clamp method [12], we were able to demonstrate the propagation of action potentials along the dendrite [13]. The waveform of an action potential recorded in the soma under the current clamp mode was reproduced in the soma by voltage clamping.

Reproduced action potential was recorded by another electrode at a dendrite of the cell. Application of TTX suppressed the amplitude of depolarization recorded at the dendrite. Thus, it is likely that the TTX-sensitive Na current in the dendrite plays a key role in the propagation of the action potential in the dendrite.

As described earlier, most amacrine cells receive GABAergic inhibitory synapse from neighboring amacrine cells. Therefore, it is likely that the GABAergic inhibition to a dendrite inhibits the propagation of the action potential only in the dendrite, leaving the propagation of the action potential from the soma to other dendrites intact. Application of GABA from a puffer pipette to a dendrite caused suppression of spontaneous action potential only in the GABA-applied dendrite, and the spontaneous action potentials in the other dendrite were unaffected [13].

How the Propagation of Depolarization Is Controlled Within an Amacrine Cell

Here, we summarize the mechanism that controls the input–output relation in a single GABAergic (possibly large field) amacrine cell. Weak excitatory inputs to a dendrite of an amacrine cell may cause a local depolarization in the dendrite, which may not propagate to the soma or to other dendrites (Fig. 1a). Under these conditions, GABA release may be local and the inhibitory output occurs only from this dendrite. If the excitatory input is large enough to activate I_{NaP} and I_{Ca}, an action potential can be evoked in the soma. The action potential will propagate into all dendrites of the cell (Fig. 1b), from which the inhibitory output will be sent out to its neighbors. GABAergic inhibition to a dendrite may suppress the propagation of the action potential from the soma only to that particular dendrite, but the action potential can still propagate into other dendrites (Fig. 1c).

In summary, I_{NaP} and I_{Ca} contribute to the propagation of the depolarizing response caused by excitatory synaptic inputs. Action potentials, once generated, convey information throughout the cell dendrites, resulting in strong inhibitory outputs. Thus, Na currents and I_{Ca} facilitate information spread within the amacrine cell to cause outputs from all dendrites of an amacrine cell. Neighboring GABAergic amacrine cells can suppress the spread of depolarization of individual dendrite independently. Therefore, the inhibitory outputs from an amacrine cell should depend on the inhibitory input to each dendrite. In other words, dendrite can act independently. Information processing in a single amacrine cell may be further elucidated by describing localization of voltage-gated channels within each morphological subtype of the amacrine cell.

Conclusions

1. Action potentials in amacrine cells cause strong inhibition in postsynaptic cells.
2. I_{NaP} and I_{Ca} enhance the excitatory synaptic inputs.
3. An action potential propagates in the dendrites of cultured GABAergic amacrine cells.
4. Propagation of the action potential in dendrites of amacrine cells can be controlled independently for each dendrite.

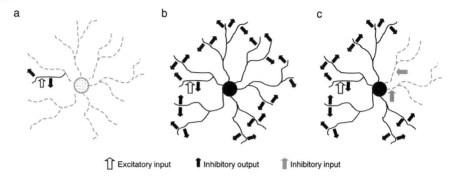

FIG. 1a–c. Control of inhibitory output triggered by an input to a dendrite of an amacrine cell.
a Weak excitatory input to a dendrite causes local inhibitory output only from the dendrite.
b A strong excitatory input activates persistent Na current (I_{NaP}) and Ca current (I_{Ca}), resulting in the generation of an action potential, which is sent to all dendrites of the cell. The inhibitory outputs are sent out to neighboring cells. **c** When some dendrites receive inhibitory inputs, action potentials in these dendrites become abortive, from which inhibitory output is not sent out, while other dendrites can still send inhibitory outputs

References

1. MacNeil MA, Masland RH (1998) Extreme diversity among amacrine cells: implication for function. Neuron 20:971–982
2. Kaneko A (1970) Physiological and morphological identification of horizontal, bipolar and amacrine cells in goldfish retina. J Physiol (Lond) 207:623–633
3. Watanabe S-I, Murakami M (1985) Electrical properties of ON-OFF transient amacrine cells in the carp retina. Neurosci Res Suppl 2:S201–S210
4. Cook PB, McReynolds JS (1998) Lateral inhibition in the inner retina is important for spatial tuning of ganglion cells. Nat Neurosci 1:714–719
5. Taylor WR (1999) TTX attenuates surround inhibition in rabbit retinal ganglion cells. Vis Neurosci 16:285–290
6. Watanabe S-I, Koizumi A, Matsunaga S, et al. (2000) GABA-mediated inhibition between amacrine cells in the goldfish retina. J Neurophysiol 84:1826–1834
7. Barnes S, Werblin F (1986) Gated currents generate single spike activity in amacrine cells of the tiger salamander retina. Proc Natl Acad Sci USA 83:1509–1512
8. Watanabe S-I, Satoh H, Koizumi A, et al. (2000) Tetrodotoxin-sensitive persistent current boosts the depolarization of retinal amacrine cells in goldfish. Neurosci Lett 278:97–100
9. Koizumi A, Watanabe S-I, Kaneko A (2001) Persistent Na$^+$ and Ca^{2+} current boost graded depolarization of rat retinal amacrine cell in culture. J Neurophysiol 86:1006–1016
10. Gleason E, Borges S, Wilson M (1994) Control of transmitter release from retinal amacrine cells by Ca^{2+} influx and efflux. Neuron 13:1109–1117
11. Miller RF, Dacheux R (1976) Dendritic and somatic spikes in mudpuppy amacrine cells: identification and TTX sensitivity. Brain Res 104:157–162
12. Stuart GJ, Sakmann B (1994) Active propagation of somatic action potentials into neocortical pyramidal cell dendrites. Nature 367:69–71
13. Yamada Y, Koizumi A, Iwasaki E, et al. (2002) Propagation of action potentials from the soma to individual dendrite of cultured rat amacrine cells is regulated by local GABA input. J Neurophysiol 87:2858–2866

Neural Mechanisms of Synchronous Oscillations in Retinal Ganglion Cells

Masao Tachibana[1], Hiroshi Ishikane[1], Mie Gangi[1], Itaru Arai[1], Tomomitsu Asaka[1], and Jiulin Du[2]

Key words. Synchronous oscillations, Retinal ganglion cell, Receptive field

Visual cortical neurons generate synchronous oscillations depending on the coherent features in the visual field. This property may serve to bind the distributed neuronal activities into a unique representation. By applying a plannar multielectrode array to the retina isolated from the frog eye, we recorded light-evoked spike discharges from multiple OFF-sustained ganglion cells (dimming detectors) and demonstrated the presence of synchronous oscillations with the correlation analysis (~30 Hz [1]). In the present study, we examined the stimulus dependence of synchronous oscillations and their neural basis. The time domain of the stimulus intensity was sinusoidally modulated at 0.25 Hz while its spatial domain was kept constant.

As the diameter of spot illumination became larger than 1.2 mm, the spike discharge rate was almost saturated but the strength of oscillations was further increased. With full-field illumination, synchronous oscillations were detected even in cell pairs more than 2 mm apart. However, when the continuity of full-field illumination was reduced, synchronous oscillations were weakened. These results suggest that synchronous oscillations among dimming detectors may serve to discriminate the size of objects, which is larger than the "classical" receptive field. The stimulus pattern that induced synchronous oscillations in dimming detectors could trigger the escape behavior in frogs.

To elucidate the neural basis of oscillatory discharges, we whole-cell current clamped dimming detectors in the flat-mount retina. During the falling phase of light stimulation, the cell showed a sustained depolarization accompanied by the rhythmical fluctuations (~30 Hz), on top of which spikes were generated. However, the injection of a depolarizing current step produced only a few spikes at its onset, and no rhythmical fluctuations were induced. The rhythmical fluctuations were not suppressed by the introduction of a Na^+ channel blocker, QX-314, into the recorded cell. However, the addition of tetrodotoxin into the bath solution completely eliminated

[1] Department of Psychology, Graduate School of Humanities and Sociology, The University of Tokyo, 7-3-1 Hongo, Bunkyo-ku, Tokyo 113-0033, Japan
[2] Shanghai Institute of Physiology, Chinese Academy of Sciences, Shanghai, China

the rhythmical fluctuations without affecting the light-evoked depolarization. The application of a gap-junction blocker suppressed the rhythmical fluctuations. These results suggest that the light-induced rhythmical fluctuations in dimming detectors are mainly originated from the synaptic inputs, which may be produced by the retinal circuit, including spiking neurons and gap junctions.

Reference

1. Ishikane H, Kawana A, Tachibana M (1999) Short- and long-range synchronous activities in dimming detectors of the frog retina. Visual Neuroscience 16:1001–1014

Adaptation of the Retinal Code: What the Eye Does Not Tell the Brain

MARKUS MEISTER

Key words. Retinal ganglion cells, Neural coding, Adaptation, Redundancy reduction, Novelty detection

The optic nerve is a bottleneck for visual information: Nowhere else along the visual pathway is the entire scene represented by such a small number of neurons. Correspondingly, the retina can transmit to the brain only a tiny fraction of the raw visual information that reaches the photoreceptors, and discards the rest. Many aspects of retinal processing can be understood as clever strategies for extracting visual features that are novel and not already available to the brain, either based on earlier transmissions, or based on other parts of the image. "Center-surround" spatial processing, and "high-pass" temporal processing are both adaptations to the average statistical structure of natural scenes and suppress the redundant components of the image. Light adaptation and contrast adaptation serve to match retinal function to changing conditions of illumination, and, again, this entails a rejection of redundant information. Recent work shows that the retina can adapt dynamically to the spatial and temporal statistics of the visual environment, even under conditions of constant mean intensity and contrast. Under many different stimulus environments, this has the effect of enhancing novel stimulus features over predictable ones. Finally, I will speculate how this general strategy of novelty detection could be implemented by modifiable circuits in the inner retina.

Department of Molecular and Cell Biology, Harvard University, Cambridge, MA 02138, USA

Coding Properties of the Avian ON-OFF Retinal Ganglion Cells

Hiroyuki Uchiyama

Key words. Retinal ganglion cell, Spike, ON-OFF interaction

Since Hartline [1] first identified three response types (ON, OFF, and ON-OFF) of vertebrate retinal ganglion cell (RGC) responses in 1938, the properties of ON and OFF RGCs have been well characterized and their function has been investigated extensively. On the other hand, it is not known how ON-OFF RGCs serve animals' vision. If spikes are evoked evenly and independently by intensity increases (ON) and decreases (OFF) in the ON-OFF RGCs, the resulting spike trains may convey only information about the timing of intensity changes of either direction, whereas more complex temporal features could be encoded if there is an asymmetrical interaction between ON- and OFF-responses [2].

In the present study, temporal properties of spike trains emitted from RGCs in response to stimuli with various temporal features were examined in urethan-anesthetized Japanese quail. Most RGCs that we sampled showed brisk response to transient stimuli. Statistical analysis of spike trains evoked by Gaussian white noise stimuli suggests that spiking time of the quail RGCs may be restrictedly regulated somehow, unlike a Poisson process. Each spike of the spike trains from ON-OFF RGCs evoked by white noise stimuli was identified as an ON- or OFF-spike, based on the slope of each spike-triggered stimulus wave. Cross-correlation analysis between those identified ON- and OFF-spikes revealed a high incidence of ON-spikes approximately 20 ms after OFF-spikes. Experiments using multiphase ramp and sequential-step stimuli confirmed the interaction. Thus, it is suggested that the OFF-machinery may affect the ON-machinery in a facilitatory manner within a short time window in the ON-OFF RGCs. Even in ON and OFF RGCs, opposite changes of light have strong inhibitory effects on responses to the stimuli to which they normally respond. Therefore, inhibitory elements interconnecting between the OFF (a) and ON (b) sublayers of the inner plexiform layer may serve for spike interaction in the ON-OFF RGCs. Thus, the ON-OFF RGCs could code temporal sequential features of stimuli with their spike timing modulated through the inner retinal circuits.

Department of Information and Computer Science, Faculty of Engineering, Kagoshima University, 1-21-40 Korimoto, Kagoshima 890-0065, Japan

References

1. Hartline HK (1938) The response of single optic nerve fibers of the vertebrate eye to illumination of the retina. Am J Physiol 121:400–415
2. Uchiyama H, Goto K, Matsunobu H (2001) ON-OFF retinal ganglion cells temporally encode OFF/ON sequence. Neural Networks 14:611–615

Spatiotemporal Changes in Gap Junctional Communication During Retinal Development and Regeneration of Adult Newt

Takehiko Saito, Hanako Oi, Yumiko Umino, Akiko Fukutomi, Chikafumi Chiba, and Osamu Kuwata

Key words. Gap junction, Connexin43, Progenitor cell

Gap junctions are one of the pathways mediating cell-to-cell communications in many biological systems. They form channels allowing ions and small molecules to pass from one cell to another. The presence of gap junctional communication in the vertebrate central nervous system (CNS) including the retina has been well documented by electrical measurements as well as dye coupling. In adult vertebrate retinas, gap junctions often form electrical synapses that mediate the rapid synaptic transmission. However, it has been also reported that gap junctions are present abundantly between cells that are not electrically excitable in the early developmental retina. Functional role(s) of such gap junctions are largely unknown.

Adult newts possess the ability to regenerate a fully functional retina from the pigment epithelium following complete removal of the neural retina. This system is attractive for understanding the mechanisms of metaplasia and cytodifferentiation. We have studied gap junctions among developing or regenerating newt retinas as models for understanding the role of gap junctional communications in the genesis of neural circuitry in the CNS. For this purpose, we prepared living slice preparations of regenerating retinas at various stages and examined changes in the degree of electrical and tracer coupling during regeneration using whole-cell patch-clamp techniques. To identify the type(s) of gap junctional proteins (Cxs), we performed immunohistochemical analyses of developing and regenerating retinas using antibodies for Cxs. We found that progenitor cells of regenerating retina couple with each other via gap junction, and that Cx43 is a major gap junctional protein between them in the early developing and regenerating retinas. We also found that the degree of electrical and tracer couplings and the amount of Cx43 gradually decreases as cellular differentiation proceeds. These results suggest that Cx43-mediated gap junctional coupling may play a role in regulating the early developmental and regenerating events, such as proliferation, cell migration, and neurogenesis.

Institute of Biological Sciences, University of Tsukuba, 1-1-1 Tennodai, Tsukuba 305-8572, Japan

Development and Plasticity in the Postnatal Mouse Retina

DAVID R. COPENHAGEN[1] and NING TIAN[1,2]

Key words. Plasticity, Development, Mouse

It has been commonly assumed that the retina has reached a nearly fully functional state by the time of eye opening. Further postnatal maturational refinement of visual function is thought to take place in higher visual areas. Moreover, the well-documented sensitivity of the visual system to visual deprivation is thought to involve exclusively reorganization of synaptic connectivity in the higher visual centers.

We report that, contrary to these generally held notions, considerable synaptic pathway reorganization occurs after eye opening in mouse and that these normally occurring maturational processes are influenced by visual activity. Using multielectrode recording arrays, we find that, just before eye opening, retinas are dominated by ganglion cells with ON-OFF light-evoked receptive fields. Two weeks after eye opening, the percentage of ON-OFF cells drop fourfold. Analysis of the relative strengths of ON and OFF pathway reveals that the loss and weakening of OFF inputs to retinal ganglion cells accounts almost exclusively for the decrease in ON-OFF responding cells. The data indicate a substantial reorganization of ON and OFF pathways following eye opening.

Dark rearing animals from before eye opening to 2 weeks afterward halts the age-related loss of OFF-responding ganglion cells. In dark-reared animals tested 2 weeks after eye opening, we find a majority of the retinal ganglion cells (RGCs) respond as ON-OFF units. The receptive field properties of the RGCs in dark-reared animals are very close to those of control-reared animals at the ages just before eye opening. These results indicate that visual activity plays an important role in the refinement of retinal circuitry.

[1] Departments of Ophthalmology and Physiology, UCSF School of Medicine, Box 0730, San Francisco, CA 94143, USA
[2] Present address: Department of Ophthalmology and Visual Sciences, Yale University School of Medicine, New Haven, CT 06520-8061, USA

Representation of Visual Information in Simple and Complex Cells in the Primary Visual Cortex

Izumi Ohzawa

Key words. Disparity-motion integration, Information multiplexing, Stereopsis

The primary visual cortex (VI) is the main gateway to visual perception. As such, neurons in this area must be able to support encoding of information relevant to many aspects of vision, including perception of form, surface, color, depth, motion, and texture. Although there is some degree of specialization, e.g., for color, within the population of neurons, the fundamental classes of cells in VI, simple and complex cells, must encode information for many aspects of visual function at the same time. If that is the case, how and to what extent do VI neurons encode form, depth, and motion information at the same time? There are relatively few existing data for addressing these questions directly, because most studies, by design, usually pay attention specifically to only one aspect of visual function at a time. We have addressed these questions by studying joint spatiotemporal receptive fields of simple cells and joint disparity-time receptive fields of complex cells in the cat striate cortex.

At the level of simple cells, a standard notion of information encoding is based on a wavelet-like image representation in which a group of simple cells tuned to a variety of orientations, spatial frequency, and phases encode a small restricted area within an image frame. Extending such an encoding scheme to the binocular case, we have previously proposed a model for binocular disparity detection by VI complex cells, known as the disparity energy model. In this model, output of a minimum of four simple cells is combined to produce a complex cell receptive field, thereby providing a concrete implementation for the hierarchical model [1]. Recent experimental data from cats, owls, and monkeys are largely consistent with predictions of the model. In addition, computational studies based on the disparity energy model show that a binocular version of the standard wavelet-like image representation can be the basis for building a hierarchical scheme for representing form and depth information within a single consistent framework.

In order to address the question of how motion information may be multiplexed with form and disparity information, we must add the time dimension to our domain

Laboratories for Neuroscience, Graduate School of Frontier Biosciences, Osaka University, 1-3 Machikaneyama, Toyonaka, Osaka 560-8531, Japan

of measurement. This is because motion is defined as change in an object's spatial position over time. For this purpose, we have measured joint spatiotemporal receptive fields for simple cells using a standard reverse correlation technique. For complex cells, we have studied special receptive fields defined in the joint domain of binocular disparity and time using m-sequence stimuli. These analyses show that simple cells may constitute subunits of a hierarchical model for complex cells that function as both motion and disparity detectors at the same time. Many complex cells possess a receptive field structure that predicts their direction selectivity well. At the same time, these neurons also possess a binocular interaction field that predicts their disparity selectivity. However, a fraction of complex cells are also found to lack disparity selectivity, exhibiting responses only to motion stimuli. In summary, we have shown that the majority of VI neurons encode information on multiple aspects of visual stimuli simultaneously in a highly multiplexed manner.

Reference

1. Hubel DH, Wiesel TN (1962) Receptive fields, binocular interactive and functional architecture in the cat's visual cortex. J Physiol Lond 160:106–154

Diversity of Computation of Orientation and Direction Preference by Visual Cortical Neurons: An Intracellular Reexamination

Yves Frégnac, Cyril Monier, Frederic Chavane, Pierre Baudot, and Lyle Graham

Key words. Synaptic excitation, Shunting inhibition, Orientation selectivity

Brain computation, in the early visual system, is often considered as a hierarchical process where features extracted in a given sensory relay are not present in previous stages of integration. In particular, many response properties in the primary visual cortex, such as orientation and directional selectivity, are not present at the preceding geniculate stage, and a classical problem is identifying the mechanisms and circuitry underlying these computations. Two main organization principles have been proposed: either the functional properties of cortical cells are defined by the feedforward drive, or they are shaped by intracortical excitatory and inhibitory architecture.

In order to assess how much synaptic integration of the network activity at each neuron contributes to the genesis of cortical orientation/direction selectivity, one must provide reliable measurements of the input/output transfer function at the single-cell level. In theory, at the level of visual cortex, a variety of combinations of excitatory and inhibitory input tuning can give rise to a given functional preference and tuning width in the spike-based response. Within this context, an important issue is whether intracortical inhibition is fundamental for stimulus selectivity, or rather normalizes spike response tuning with respect to other features such as stimulus strength or contrast, without influencing the selectivity bias and preference expressed in the excitatory input alone. After a long-standing debate, most recent models favored the second possibility, in large part because experimental support for the existence of inhibitory input in response to nonpreferred stimuli has been somewhat contradictory.

Faced with the diversity of the experimental observations concerning the presence or not of inhibitory input evoked by nonpreferred orientation/directions, we decided to readdress this issue with a quantitative comparative study combining both sharp and patch recordings. Visually evoked synaptic activity was measured intracellularly in simple and complex cells in current clamp and voltage clamp. Different stages of

Department of Integrative and Computational Neurosciences, UNIC UPR CNRS2191, Gif-sur-Yvette, France

postsynaptic integration were compared on the basis of changes in the spiking rate, the subthreshold membrane potential mean, and the stimulus-locked variability evoked from the resting state. Two additional methods were used to quantify input tuning by increasing the driving force of inhibition during spike inactivation, and by directly monitoring excitatory and inhibitory conductances. Results suggest three major patterns of interaction. In a first scheme (P-P), excitatory and inhibitory inputs demonstrate the same preference as the spike. In a second scheme (P-NP), inhibition is the strongest for nonpreferred stimuli. In the last scheme (NP-NP), the excitatory and inhibitory inputs are both optimal for nonpreferred stimuli.

We propose that the diversity of input combinations found across cells reflects anatomical nonhomogeneities in the lateral intracortical connectivity pattern whose functional impact may be up- and down-regulated by correlation-based activity-dependent processes.

Acknowledgments. This work was supported by grants from the CNRS (Bioinformatics) and HFSP RG-103-98 to Y.F.

Assembly of Receptive Fields in Cat Visual Cortex

DAVID L. FERSTER

Key words. Visual cortex, Orientation selectivity, Intracellular recording

The origin of orientation selectivity in the responses of simple cells in cat visual cortex serves as a model problem for understanding cortical circuitry and computation. The feedforward model of Hubel and Wiesel [1] posits that this selectivity arises simply from the spatial organization of the receptive fields of thalamic inputs synapsing on each simple cell. Much evidence, including a number of recent intracellular studies, supports a primary role of the thalamic inputs in determining simple-cell response properties including orientation tuning. And yet, while the feedforward model seems to explain the broad outline of simple-cell properties, there are number of detailed aspects of the behavior of simple cells that have appeared not to be accounted for by the feedforward model. These properties include contrast invariance of orientation tuning, the exact relationship between receptive field geometry and orientation tuning, and the dynamics of orientation tuning. The apparent failures of the feed-forward model have prompted the development of a class of models that rely on feedback circuitry within the cortex: Properly arranged feedback from excitatory connections within an orientation column and inhibitory connections to adjacent columns can account for many of the properties that the feedforward models miss. The feedforward and feedback models are different enough in character that they make radically different predictions about the nature of the computation performed by the cortex, the feedforward models acting more like passive filters, and the feed-back models more actively shaping the representation of the retinal image. I will review a series of experiments designed to test these two models, and present evidence that the feedforward models, with little modification, can account in detail for the behavior of cortical simple cells.

Reference

1. Hubel DH, Wiesel TN (1962) Receptive fields, binocular interaction and functional architecture in the cat's visual cortex. J Physiol 160:106–154

Department of Neurobiology and Physiology, Northwestern University, 2205 Tech Drive, Evanston, IL 60202, USA

Local Circuit in the Cerebral Cortex

TAKESHI KANEKO

Key words. Local circuit, Pyramidal neuron, Cerebral cortex

To reveal the mechanism of information processing in the cerebral cortex, we focused on the morphological analysis of its intrinsic circuitry. Recently, we developed a retrograde neuronal tracing method with Golgi-like soma-dendritic filling. By combining this method with the intracellular staining method, the local connection from pyramidal cells to corticospinal and corticothalamic projection neurons were examined in the motor and somatosensory cortical areas of the rat.

Layer III pyramidal neurons stained intracellularly were of the regular spiking type, showed immunoreactivity for glutaminase, and emitted axon collaterals arborizing locally in layers II/III and/or V. Nine of them were reconstructed for morphological analysis; 15.2% or 3.8% of varicosities of axon collaterals of the reconstructed neurons were apposed to dendrites of corticospinal or corticothalamic neurons, respectively. By con-focal laser scanning and electron microscopy, some of these appositions were revealed to make synapses. These findings indicate that corticospinal neurons receive information from the superficial cortical layers fourfold more efficiently than corticothalamic neurons. The connections were further examined by intracellular recording of excitatory postsynaptic potentials that were evoked in layer V and layer VI pyramidal neurons by stimulation of layer II/III. Most layer V pyramidal cells received monosynaptic inputs, whereas layer VI pyramidal cells admitted polysynaptic inputs from layer II/III.

By using a similar technique, we investigated the axon collateral connections to corticospinal projection neurons in layer V of the motor and motor sensory areas. Corticospinal neurons generally received axon collateral inputs from pyramidal cells of all cortical layers except layer I. These results indicate that corticospinal neurons in layer V collect information on all the cortical layers, especially processed information of the superficial cortical layers, and make a command to the subcortical regions.

In contrast to corticospinal neurons, corticothalamic projection neurons in layer VI of the somatosensory cortex received many more axon collateral inputs from pyra-

Department of Morphological Brain Science, Graduate School of Medicine, Kyoto University, Yoshida, Sakyo-ku, Kyoto 606-8501, Japan

midal cells in layers IV and VI than did those in layer II/III and layer V. Thus, corticothalamic neurons are considered relatively independent of or only poorly related to information processing of the superficial cortical layers or command output of layer V neurons. Furthermore, since layers IV and VI directly receive thalamic inputs, it is likely that corticothalamic neurons, together with thalamocortical neurons, exert as a simple dynamical system, which is relatively independent of information processing in the superficial cortical layers.

Properties and Possible Mechanisms of Response Modulation by the Receptive Field Surround in the Cat Primary Visual Cortex

Hiromichi Sato

Key words. Primary visual cortex, Surround suppression, Excitation and inhibition

In neurons of the primary visual cortex (V1), the response to the grating stimulus presented in the classical receptive field (CRF) is suppressively modulated by the grating presented at the receptive field surround (SRF). This surround suppression (SS) depends on the relationship of stimulus parameters, such as orientation and spatial frequency, between the CRF and SRF gratings. The strength of SS depends on the stimulus contrast of both CRF and SRF. To assess the input mechanism underlying this stimulus-context-dependent suppression, we performed two series of experiments on the contribution of intracortical inhibition to SS and the nature of the contrast-gain control mechanism of SS.

First, we tested the effects of an iontophoretically administered γ-aminobutyric acid$(GABA)_A$-receptor antagonist, bicuculline methiodide (BMI), on the responses of 45 V1 neurons in anesthetized and paralyzed cats. The tuning curves of response to grating patches of various sizes with or without BMI were compared in each cell. When BMI was administered, the mean firing rate of the cells increased, but the shape of the normalized tuning curve did not change, suggesting that the intracortical inhibition does not play a primary role in SS. Moreover, the spatial distributions of excitatory and inhibitory summation field were estimated to be essentially similar to each other.

Second, we tested the suppressive effects of SRF gratings at varying contrast on the responses to the CRF stimulation using either a high- or low-contrast grating. That is, the suppressive surround effect was tested at different saturation levels of CRF responses. In 40 cells that exhibited strong SS, the SRF grating progressively suppressed the CRF responses as the degree of the grating contrast increased, regardless of the contrast of the CRF grating. Basically, the surround-contrast-response curves that express the relationship between contrast of the SRF grating and the response magnitude exhibited quite similar patterns in terms of shape for different contrasts of CRF stimuli. The surround-contrast-response curves suggested the presence of response-gain control and contrast-gain control mechanisms for the suppressive response modulation by the SRF stimulus.

School of Health and Sport Sciences and Graduate School of Frontier Biosciences, Osaka University, 1-17 Machikaneyama, Toyonaka, Osaka 560-0043, Japan

Representation of Surface Luminance and Brightness in Macaque Visual Cortex

Hidehiko Komatsu[1] and Masaharu Kinoshita[1,2]

Key words. VI, Surface, Brightness

Perceived brightness of a surface is determined not only by the luminance of the surface (local information), but also by the luminance of its surround (global information). Recent studies have indicated that activities of neurons in the early visual areas are affected not only by stimuli within their receptive fields (RFs), but also by stimuli outside the RF. This suggests that the interaction between local and global information in brightness perception may be accounted for in terms of contextual modulation of the activities of luminance-sensitive neurons in the early visual area. We investigated the effects of local and global luminance on the activity of neurons in the primary visual cortex (VI) of awake macaque monkeys performing a visual fixation task. Neural responses to homogeneous surface stimuli at least three times as large as the receptive field were examined. We first tested the sensitivity of neurons to variation in local surface luminance while the luminance of the surround was kept constant. A large majority of surface-responsive neurons (106/115) changed their activity depending on the surface luminance in monotonic fashion; this was more evident in the later response period. Then, the effect of the global luminance was examined in 81 neurons by varying the luminance of the surround while keeping the surface luminance constant. The responses of some neurons (25/81) were not affected by the surround luminance. In other neurons, responses were affected by the luminance of the surround. In some of these neurons, the effects of the luminance of the surface and the surround were in the same direction, while in the others, the directions were the opposite. The former seem to encode the level of illumination, while the responses of the latter parallel the perceived brightness of the surface. These findings suggest that significant interaction between local and global luminance information takes place in VI and perceived brightness of a homogeneous surface is represented at the level of VI.

[1] Laboratory of Sensory and Congenitive Information, National Institute for Physiological Sciences, Myodaiji, Okazaki, Aichi 444-8585, Japan
[2] Present address: The Rockefeller University, 1230 York Avenue, New York, NY 10021, USA

Inhibitory Circuits that Control Critical Period Plasticity in Developing Visual Cortex

Takao K. Hensch

Key words. Critical period, GABA, Ocular dominance

In the visual system, input from the two eyes is typically well balanced, but occluding one during a brief time in early life can permanently rewire the brain in favor of the other. We have found this sensitivity to emerge upon maturation of a particular class of inhibitory connection within neocortex, identifying a mechanism that triggers the "critical period" for plasticity [1].

Early in postnatal life, weak γ-aminobutyric acid (GABA)ergic transmission in visual cortex produces prolonged neuronal discharge and an inability to discriminate imbalanced sensory input. Both can be rescued by the allosteric $GABA_A$ receptor modulator diazepam, leading to premature expression of the critical period. Conversely, in mice genetically engineered to maintain low levels of synaptic inhibition (GAD65 knock-out), the critical period is never activated unless inhibition is pharmacologically enhanced for as little as 2 days. Once triggered, sensitivity to visual manipulation dramatically disappears in both wild-type and mutant animals.

Among the large heterogeneity of neocortical inhibitory interneurons, chandelier cells represent a unique example of synapse specificity, the main source of input onto pyramidal cell axon initial segments, where $GABA_A$ receptor α2-subunits are preferentially localized. We have directly tested the role of these connections, using mice whose α-subunits were selectively rendered insensitive to benzodiazepines by knock-in of a point mutation (H101R). While ideally suited for controlling prolonged discharge, other GABAergic synapses utilizing α1-subunits were found to drive ocular dominance plasticity. Large basket cells, a class of parvalbumin-positive interneuron regulated by age and sensory experience, are an attractive candidate.

Forty years after Hubel and Wiesel [2], microdissection of visual cortical plasticity is now possible based upon inhibitory circuits. Dissociation from aberrant neural coding carries broad implications for models of brain development and function. Potential downstream molecular cascades are further pointing us toward a true cellular understanding of critical period plasticity in the neocortex.

Laboratory for Neuronal Circuit Development, RIKEN Brain Science Institute, 2-1 Hirosawa, Wako, Saitama 351-0198, Japan

References

1. Fagiolini M, Hensch TK (2000) Inhibitory threshold for critical-period activation in primary visual cortex. Nature 404:183–186
2. Hubel DH, Wiesel TN (1962) Receptive fields, binocular interaction and functional architecture in the cat's visual cortex. J Physiol (Lond) 160:106–154

Substrates of Rapid Plasticity in Developing Visual Cortex

Michael P. Stryker

Key words. Plasticity, Area 17, LGN

The loss of visual responses by neurons in the primary visual cortex to an eye in which vision is occluded at the peak of the critical period is very rapid, appearing from extracellular microelectrode recordings to be nearly maximal after 2 days of monocular deprivation (MD). Anatomically, the first changes in thalamocortical afferent inputs to layer IV appear after 4 days of MD in the cat, and 6–7 days of MD are needed for near maximal losses of deprived-eye input. When we examined the different cortical layers, rapid plasticity was present in all except layer IV, removing the apparent mismatch between anatomical and physiological measures of plasticity in that layer.

Layer III represents a second stage of cortical processing beyond the input layer, and it shows very rapid plasticity in response to MD or strabismus at the peak of the critical period. We considered two possible anatomical substrates of this rapid plasticity: ascending connections from layer IV cells to initially binocularly driven layer III neurons, and tangential connections from the different ocular dominance columns within layer III. The former studies are incomplete, but the latter revealed a greater than 20-fold change in the density of tangential connections after only 2 days of visual deprivation by strabismus. Anatomical connections to the centers of other ocular dominance columns serving the same eye were increased more than threefold, and connections to the centers of opposite-eye ocular dominance columns were decreased more than sixfold. These findings show that the anatomical effects of visual deprivation are fully as rapid and great in magnitude as any physiological changes that take place in this system.

Additional studies were carried out in mice to determine whether any measurable component of physiological plasticity could take place without the protein synthesis that is necessary for these anatomical changes. Protein synthesis inhibition in the visual cortex completely blocked any physiological indication of plasticity, while rapid plasticity was entirely normal following inhibition of geniculate protein synthesis.

Together these experiments indicate that cortical plasticity occurs by rapid anatomical changes over hours to days in the upper layers followed by thalamocortical afferent rearrangements in the week-to-weeks time scale.

Department of Physiology, UCSF, 513 Parnassus Avenue, Room S-762, San Francisco, CA 94143, USA

Actions of Brain-Derived Neurotrophic Factor on Function and Morphology of Visual Cortical Neurons

Tadaharu Tsumoto, Naoki Takada, and Keigo Kohara

Key words. Neurotrophin, Glutamate receptor, Visual cortex

Brain-derived neurotrophic factor (BDNF) is known to play a role in experience-dependent plasticity of the developing visual cortex. For example, BDNF acutely enhances long-term potentiation and blocks long-term depression in the visual cortex of young rats [1–3]. Such acute actions of BDNF are suggested to be mediated mainly through presynaptic mechanisms. A chronic application of BDNF to the visual cortex of kittens is known to expand ocular dominance columns in the cortex [4]. Recently we have demonstrated that BDNF is transferred from presynaptic axon terminals to postsynaptic neurons in an activity-dependent manner, suggesting the possibility that BDNF also has chronic postsynaptic actions [5]. To test this possibility, we carried out two types of experiments.

First, we examined whether the chronic application of BDNF has any effect on postsynaptic receptor function in solitary neurons cultured from visual cortex of BDNF knockout and wild-type mice. BDNF at the concentration of 100 ng/ml was applied to one of the two groups of neurons cultured from each type of mice for 8–11 days, and then excitatory postsynaptic currents (EPSCs) were recorded using the whole-cell voltage-clamp technique. We found that the amplitude of both types of glutamate receptor-mediated components, N-methyl-D-aspartate (NMDA) and α-amino-3-hydroxy-5-methyl-4-isoxazole propionate (AMPA) components of evoked EPSCs recorded from neurons of the wild type was larger than that from neurons of the BDNF knockout type. We also found that chronic treatment with BDNF enhanced the amplitude of both components in either type of neuron. In addition, the degree of enhancement in NMDA receptor-mediated EPSCs by BDNF was larger than that in AMPA receptor-mediated EPSCs in either type of neuron. These results indicate that the chronic application of BDNF to cultured cortical neurons preferentially enhances NMDA receptor-mediated responses.

Second, we tested whether endogenous BDNF has any effect on dendritic morphology of postsynaptic neurons in "chimera culture" of visual cortical neurons pre-

CREST/Japan Science and Technology Corporation, and Division of Neurophysiology, Osaka University Graduate School of Medicine, 2-2 Yamadaoka, Suita 565-0871, Japan

pared from two types of transgenic mice, green fluorescence protein (GFP) mice and BDNF knockout mice. Neurons derived from the former mice have endogenous BDNF as well as GFP: We found that neurons derived from the latter mice, BDNF (−/−) neurons, had relatively poor dendrites if they were not contacted by GFP-positive terminals, whereas BDNF (−/−) neurons had complex dendritic morphology if they were directly contacted by GFP-positive terminals and thus supplied with endogenous BDNF. These results indicate that endogenous BDNF plays a role in the development of dendrites of visual cortical neurons.

References

1. Akaneya Y, Tsumoto T, Hatanaka H (1996) Long-term depression blocked by brain-derived neurotrophic factor in rat visual cortex. J Neurophysiol 76:4198–4201
2. Akaneya Y, Tsumoto T, Kinoshita S, et al. (1997) Brain-derived neurotrophic factor enhances long-term potentiation in rat visual cortex. J Neurosci 17:6707–6716
3. McAllister AK, Katz LC, Lo DC (1999) Neurotrophins and synaptic plasticity. Annu Rev Neurosci 22:295–318
4. Hata Y, Ohshima M, Ichisaka S, et al. (2000) BDNF expands ocular dominance columns in visual cortex in monocularly deprived and non-deprived kittens, but not in adult cats. J Neurosci 20RC57:1–5
5. Kohara K, Kitamura A, Morishima M, et al. (2001) Activity-dependent transfer of brain-derived neurotrophic factor to postsynaptic neurons. Science 291:2419–2423

Distance and Size Estimation in the Tiger Beetle Larva: Behavioral, Morphological, and Electrophysiological Approaches

Yoshihiro Toh, Jun-ya Okamura, and Yuki Takeda

Key words. Tiger beetle, Stemma, Visual interneuron, Optic neuropil, Vision

Introduction

Estimation of the size and distance of an object is one of the important visual functions for a wide variety of animals. How do animals visually measure distance and size? The size of the retinal image of a given object changes as a function of both distance and its absolute size. Since higher animals can learn the absolute size of many objects in their visual worlds, they can estimate the distance to a particular object by its apparent angular size from knowledge of its absolute size gained by memory. Of course, humans can also estimate an object's distance by binocular disparity.

Insects also estimate the absolute size and distance of an object. They can, for example, fly well without colliding with various obstacles. A praying mantis steals up on a prey insect until it can strike the prey by its foreleg in a grass field. It is, however, unlikely that insects estimate an object distance based upon the absolute size and apparent angular size, because their brains do not seem to be large enough to store memories of the absolute sizes of many articles. Motion parallax is more widely accepted as a mechanism for distance estimation in insects, although binocular cues are also available in some insects (e.g., [1]).

Even though binocular vision, motion parallax, and other mechanisms have all been proposed for distance estimation in insects, the underlying mechanisms are not fully understood at a neural level. Although visual interneurons tuned to small objects have been reported in several insects, they might respond to apparent angular size but not to absolute size (e.g., [2]).

The tiger beetle larva shows distinct visual behavior to moving objects [3]. The larva is an ambushing hunter with a body length of 15–22 mm that lives in a tunnel in the ground. It ambushes prey, keeping its head horizontal at the opening of the tunnel. When the prey approaches the tunnel, the larva jumps to snap at it, but if a large object moves within the jumping range (about 15 mm) the larva quickly with-

Department of Biology, Graduate School of Science, Kyushu University, 6-10-1 Hakozaki, Higashi-ku, Fukuoka 812-8581, Japan

draws deep into the tunnel. When an object moves beyond its jumping range, the larva always shows the escape response, regardless of the object's size. These responses are mediated by input from only two of six pairs of stemmata. Previous work suggested that both binocular vision and monocular vision were concerned with the estimation of the distance [3]. It was shown that some visual interneurons have distance sensitivity, but the morphology of these cells was not identified [4]. In the present study, the responses and morphology of visual interneurons have been examined with reference to distance sensitivity and size preference.

Materials and Methods

Third (final) instar larvae of the tiger beetle, *Cicindela chinensis*, were used in the present study. The larva was waxed to an acrylic specimen holder for intracellular recording. The pair of optic lobes was exposed, and a glass microelectrode filled with Lucifer yellow was advanced into the medulla (the second optic neuropil). After recording, the impaled neuron was intracellularly stained and examined by confocal scanning microscopy. The apparent motion of a black square on a liquid crystal display placed at a distance from 10 to 50 mm from the larva was used for visual stimulus. Details of the methods are dealt with elsewhere [4].

Results and Discussion

Structure of Visual System

The structure of the visual system of the tiger beetle larva has been dealt with elsewhere [5–7], but will be briefly described here. The larva possesses six stemmata on either side of the head, two of which are much larger than the others and referred to as the anterior and posterior large stemmata (Fig. 1A,B). A flattened heart-shaped optic neuropil complex occurs beneath the six stemmata. The anterior half of the neuropil complex is that of the anterior large stemma, whereas the posterior half is that of the posterior one (Fig. 1C). The distal region (lamina) of the complex is the first optic neuropil and the proximal region (medulla) is the second optic neuropil. The lamina neuropil is subdivided for the two large stemmata, whereas the medulla neuropil is not clearly subdivided. The medulla neuropil is connected to the brain by an optic nerve, which contains approximately 120 axons.

Distance-Sensitive Neurons

Okamura and Toh [4] reported that a few medulla neurons showed distance sensitivity, whereas the majority of medulla neurons were insensitive to object distances. Although the morphology of some medulla neurons was shown after recording [8], the relation of these to distance preference was not examined. In the present study, distance-sensitive neurons were intracellularly stained. A neuron shown in Fig. 1D responds to a moving object at a distance of 10 mm, but not to one that has a distance of 50 mm (Fig. 2). The neuron extends two large dendritic branches, which terminate beneath the peripheral parts of the lamina neuropils of the two large stemmata (Fig.

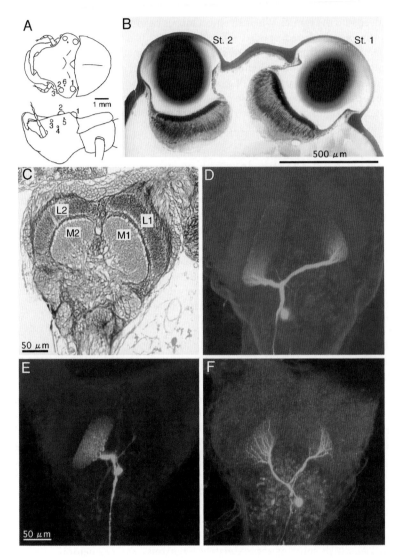

FIG. 1. **A** Distribution of six stemmata (*1–6*) on the head of the tiger beetle larva. Dorsal view (*upper*) and lateral view (*lower*). **B** Sagittal section through two large stemmata. *St. 1*, posterior large stemma; *St. 2*, anterior large stemma. **C** Horizontal section of optic neuropil complex. *L1*, lamina neuropil of the posterior stemma; *L2*, lamina neuropil of the anterior stemma; *M1*, medulla neuropil of the posterior stemma; *M2*, medulla neuropil of the anterior stemma. **D–F** Lucifer yellow-filled medulla neurons. Responses of neurons in **D**, **E**, and **F** are shown in Figs. 2, 3A, and 3B, respectively

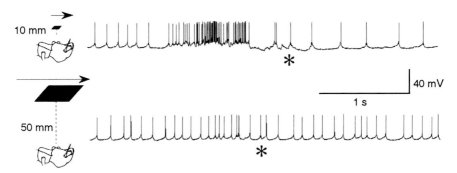

FIG. 2. Responses of a medulla neuron to black squares (11°) moving at distances of either 10 mm or 50 mm. This neuron responds with an increased spike discharge rate to a black square moving at a distance of 10 mm (*upper*), but does not respond to one moving at a distance of 50 mm (*lower*). *Asterisks* indicate when the black square passes just above the head of the larva.

1D). Since the lamina neuropil is organized retinotopically, it seems that the recorded neuron must receive inputs from the peripheral part of the retina in the two large stemmata. Such morphology seems to favor the idea proposed by Toh and Okamura [9], as follows. On the basis of the different distances from the lens to the retina between the central and the peripheral parts of the retina in the large stemma, they proposed the possibility that a close object might form an image in focus in the peripheral retina, whereas a distant object forms an in-focus image in the central part of the retina. If we suppose that only a clearly focused image is encoded by follower neurons, a close object may be perceived in the peripheral part of the retina. However, this idea appears too simple and has some difficulties. Its validity should be tested by morphological comparisons between distance-sensitive and distance-insensitive neurons.

Preference to Angular Size

When they moved at distances less than 15 mm, the larvae jumped to small objects but escaped from large objects. These data suggest that some neurons in the visual system retain size preference, but their responses to objects of different sizes have not been examined. In the present study, the responses of medulla neurons to apparent movement of either a small black square (3 mm) or a large one (20 mm) at a distance of 10 mm were examined. Although many different response patterns were recorded, two are shown in Fig. 3. The neuron in Fig. 3A responds only to a small black square, but not to a large one. This neuron has a narrow dendritic arborization beneath the lamina neuropil of the anterior large stemma (Fig. 1E). The neuron in Fig. 3B responds only to a large black square. This neuron has two thick dendrites, arborizations of which are widely spread beneath the lamina neuropils of the two large stemmata (Fig. 1F). These neurons together with other types of neurons seem to play a role in the estimation of object size.

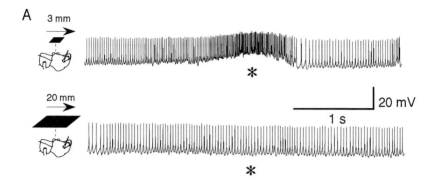

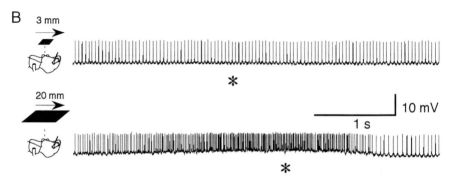

FIG. 3A,B. Responses of two medulla neurons to black squares of 3 mm and 20 mm moving at a distance of 10 mm. **A** This neuron responds with increased spike discharges to a square of 3 mm (*upper*), but not to one of 20 mm (*lower*). **B** This neuron responds with increased spike discharges to a square of 20 mm (*lower*), but not to one of 3 mm (*upper*). *Asterisks* indicate when the black square passes just above the head of the larva.

Summary and Further Problems

In the present study, both distance-sensitive neurons and neurons signaling a size preference have been identified in the visual system of the tiger beetle larva. However, a peripheral mechanism for how such response patterns are established and the central projections by means of which they are integrated in the brain both still remain unknown.

Acknowledgments. The authors express their thanks to Professor I.A. Meinertzhagen (Dalhousie University) for critical revision of the manuscript. This work was supported in part by a Grant-in-Aid for Scientific Research from the Ministry of Education, Science, and Culture of Japan 60037265.

References

1. Schwind R (1989) Size and distance perception in compound eyes. In: Stavenga DG, Hardie RC (eds) Facets of vision. Springer, Berlin, Heidelberg, New York, pp 425–444
2. Egelhaaf M (1985) On the neuronal basis of figure-ground discrimination by relative motion in the visual system of the fly. Biol Cybern 52:195–209
3. Mizutani A, Toh Y (1998) Behavioral analysis of two distinct visual responses in the larva of the tiger beetle (*Cicindela chinensis*). J Comp Physiol 182:277–286
4. Okamura J-Y, Toh Y (2001) Responses of medulla neurons to illumination and movement stimuli in the tiger beetle larvae. J Comp Physiol 187:712–725
5. Friederichs HF (1931) Beitrage zur Morphologie und Physiologie der Sehorgane der Cicindelinen. Z Morph Oekol Tiere 21:1–172
6. Toh Y, Mizutani A (1994a) Structure of the visual system of the larva of the tiger beetle (*Cicindela chinensis*). Cell Tissue Res 278:125–134
7. Toh Y, Mizutani A (1994b) Neural organization of the lamina neuropil of the larva of the tiger beetle (*Cicindela chinensis*). Cell Tissue Res 278:135–144
8. Mizutani A, Toh Y (1995) Optical and physiological properties of the larval visual system of the tiger beetle, *Cicindela chinensis*. J Comp Physiol 178:591–599
9. Toh Y, Okamura J-Y (2001) Behavioural responses of the tiger beetle larva to moving objects: role of binocular and monocular vision. J Exp Biol 204:1–12

Extraocular Photoreception of a Marine Gastropoda, *Onchidium*: Three-Dimensional Analysis of the Axons of Dermal Photoreceptor Cells in the Dorsal Mantle Examined with a High-Voltage Electron Microscope

Nobuko Katagiri[1], Yuichi Simatani[2], Tatsuo Arii[3], and Yasuo Katagiri[4]

Key words. Extraocular photoreception, *Onchidium*, Dermal photoreceptor cell, Rhabdomeric-type photoreceptor cell

Introduction

Extraocular photoreception exists in a variety of forms among invertebrates [1], and cannot be identified visually from a surface view. Messenger [2] defined extraocular photoreception as, "a response to light that is not mediated by an eye." Dermal light sense is one type of extraocular photoreception. In molluscs, extraocular photo-receptors are widespread and exhibit a surprising range of complexity. The shadow responses, categorized as dermal sensitivity by Messenger [2], are widespread in both gastropods and bivalves and are mediated by cells or sensory endings that remain unidentified. The marine gastropod *Onchidium* is thought to have extraocular pho-toreception based on behavioral responses to light and shadow [1–3]. *Onchidium* possesses paired stalk eyes (SEs) and dorsal eyes (DEs), demonstrating that both dermal and ocular systems can coexist. Yoshida [1] has described the dominant role the dermal system plays in *Onchidium*. However, the dermal light sensitivity of *Onchidium* should be reconsidered, as a photoreceptor cell has been found in the dorsal mantle [3, 4].

[1] Medical Research Institute, Tokyo Women's Medical University, 8-1 Kawada-cho, Shinjuku-ku, Tokyo 162-8666, Japan
[2] Department of Physiology, Tokyo Women's Medical University School of Medicine, 8-1 Kawada-cho, Shinjuku-ku, Tokyo 162-8666, Japan
[3] National Institute for Physiological Sciences, 38 Saigo-naka, Myodaiji-cho, Okazaki, Aichi 444-8585, Japan
[4] Section of Basic Science, Tokyo Women's Medical University School of Nursing, 400-2 Shimohijikata, Daito-cho, Shizuoka 437-1434, Japan

The Multiple Photoreceptive System of Onchidium

Onchidium has a unique multiple photoreceptive system comprising SEs, DEs, photosensitive ganglion cells in the central nervous system (CNS), and extraocular photoreceptors [3–5]. The SE is situated at the tentacle and the DEs are located on the ocellus papillae of the dorsal mantle. Many extraocular photoreceptors are widespread on the dorsal mantle, but cannot be detected because of the lack of dark pigment. The giant cells described by Stantschinsky [6] are distributed broadly in the connective tissue of various-sized papillae on the dorsal mantle of *Onchidium*. On the basis of their fine structure and photoresponse, we named these giant cells the dermal photoreceptor cell (DPC) [3].

Dermal Photoreceptor Cell

Fine Structure

DPCs are distributed independently or in small clusters without special capsular structure in the dermis. They are generally oval in shape and vary in diameter from 40 to 100 µm. The distal third of the cell consists of massive microvilli and the proximal cytoplasm has a large nucleus with maze-like incision. A great number of photic vesicles of uniform size (80 nm) were packed closely in the cytoplasm and became black with prolonged osmium impregnation (Fig. 1). No synaptic structure was found on the DPC body or on the DP axon.

The fine structure of the DPC is typical of the rhabdomeric-type photoreceptor cells of gastropod eyes, which include the SE of *Onchidium* that contains microvilli and photic vesicles [7, 8]. Photic vesicles are a special form of endoplasmic reticulum. Both organelles become black upon osmium impregnation [7, 9]. The DPC is also similar to lens cells of the DE except for photic vesicles, and is similar to the ganglion cells of the retina in *Onchidium* SE, which is a presumed circadian pacemaker [8]. The lens cell contains an abundance of endoplasmic reticulum in the cytoplasm [3]. The DPC is an intermediate structure between the type 1 visual cells of the SE [8] and the lens cells of the DE.

The DPC is an extraocular photoreceptor that is less organized than the eye. The widespread distribution of the DPC over the entire dorsal mantle allows the detection of sudden, undefined photic information, indicating that the DPC may participate in the peripheral reflex system.

Axon of DPC

The single axon of the DPC extends from the confined lateral side of the DPC body, adjacent to the basal area of microvilli, and meanders through connective tissue, finally joining a small papillar nerve (SPN) located outside the DPC cluster [4] (Fig. 2). The axon is variable in size and structure along its length and is enclosed within a sheath. The DPC of an osmium-impregnated juvenile was examined and reconstructed three-dimensionally on high-voltage transmission electron microscope micrographs of serial semi-thin sections using the OZ-95-32 3D Reconstruct System (Rise, Sendai, Japan). An axon was observed to arise from the lateral side, run together with the axon of a neighboring DPC, and join a small nerve bundle, as seen in adult

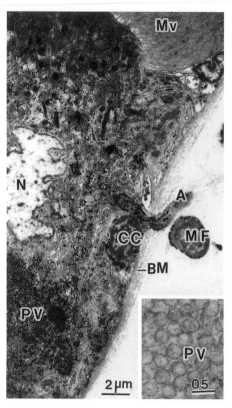

2 μm

05

FIG. 1. The axon (*A*) extends directly from the lateral side of the dermal photoreceptor cell (DPC). The surface of the DPC is covered with supporting cells (*CC*). Photic vesicles (*PV*) in the cytoplasm were clear in the ordinarily fixed DPC (*insert*), but were strongly impregnated with osmium. *BM*, basal lamina; *MF*, muscle fiber; *Mv*, microvilli; *N*, nucleus

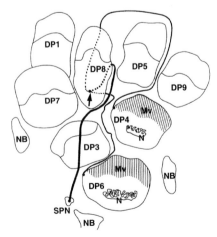

FIG. 2. Two dermal photoreceptor cells (DPCs), DP6 and DP4, are located in a group of 19 DPCs. The axons of DP6 and DP4 run separately from the site of emergence to a point (*arrow*) behind DP8, then run parallel to the site joining the small papillar nerve (*SPN*). Axons were observed through 190 of 700 serial semi-thin sections (0.4 μm) of osmium-impregnated adult specimen, at 1000 kV with an H-1250M Hitachi electron microscope (National Institute for Physiological Science). Tracing the axons and projecting the position of axons in each section schematically onto a micrograph enabled the three-dimensional reconstruction of the entire course of DP6 and DP4 axons. *Mv*, microvilli; *N*, nucleus, *NB*, nerve bundle

DPCs. In many species the axon usually extends from the basal portion of the photoreceptor cell. The unique emergence of the DPC axon has been reported in the distal cell of the scallop eye [7] and in the ganglion cell such as *Onchidium* SE [8]. The route of these axons, from emergence to the joining site into the SPN, has not been reported in other species.

Two Cell Types Associated with the Axon

Supporting cells cover the surface of the DPC, including the initial portion of the axon (Fig. 1), and sheath cells enclose the SPN and axon continuously. These two cell types may participate in the sheath structure of the axon, as no other cell types follow the course of the axon.

Glial cells cover the surface of earthworm photoreceptor cells, which are similar in distribution and fine structure to DPCs [10]. Glial cells of invertebrates wrap the axons either singly or in groups to speed up impulse conduction [11]. DP axons may be surrounded by a single glial cell sheath, whereas the SPN is sheathed by a group of glial cells.

Photoresponse of the DPC

The DPC showed resting membrane potentials between −40 and −70 mV in the dark. The depolarizing response to light was recorded by intracellular electrodes and was associated with an increase in membrane conductance. Spectral sensitivity peaked at 520 nm. The spike response to light, on-discharge, was recorded by suction electrodes from the papillar nerve of isolated papilla containing DPCs [3].

Smooth Muscle Cells (SMCs)

SMCs are distributed randomly in the connective tissue. In juveniles, we found giant stellate SMCs similar to ordinary SMCs in fine structure, but possessing an irregular-shaped perikaryon and multiple long processes extending in various directions. Long processes of stellate SMCs branch and terminate at the basal lamina of epidermal cells. Together, the giant stellate SMCs form a basket-like network with their long processes and enclose the DPCs and nerve bundle.

Stellate SMCs have not been identified in the adult papillae. This may be because of difficulty in identifying stellate SMCs because of their large size and wide distribution.

Shadow Reflex of the Dorsal Mantle

A creeping *Onchidium* extends its tentacles, containing SEs, toward the proceeding direction, and the ocellus papillae, containing DEs on the dorsal mantle, are extended upwards. When a small shadow falls on the dorsal surface of the mantle, only the local area responds by rapidly indenting. The shadow response remained after complete resection of SEs and DEs. Isolated papillae from the mantle showed a local contraction response to light.

It is possible that the shadow reflex system consists of the photoreceptor, effectors, and nerve fibers in the *Onchidium* dorsal mantle. The DPC may be a photoreceptor, and the stellate SMCs may function as effectors for contraction. However, it is unknown how information passes between the DPC and the effector SMCs.

Pathway of Photic Information

The SPN contains several DPC axons and unidentified nerve fibers and forms a thick nerve bundle (NB), together with other SPNs of the same and different papillae. The pleural nerve trunk innervates the dorsal mantle, and its branches terminate in the DEs and the epithelial sense organ (DPCs). Thicker SPN may connect to the pleural nerve trunk, although the connection between them was not visible, as only small papillae were examined. The DPC axon may communicate photic information from the DPC to the CNS, or to an unidentified nerve fiber, via synaptic structures in the SPN. This nerve fiber may terminate in the SMCs, leading to contraction and a local shadow reflex.

Yoshida [1] classified the shadow response of *Onchidium* as "dermal photosensitivity." Now that the fine structure and photoresponse of the DPC have been elucidated [3, 4], dermal photoreception of the *Onchidium* dorsal mantle is more appropriately classified as "unorganized photoreceptive cells" in "diffuse photosensitivity."

References

1. Yoshida M (1979) Extraocular photoreception. In: Autrum H (ed) Handbook of sensory physiology, vol. 7/6A: A. Invertebrate photoreceptors. Springer-Verlag, Berlin, Heidelberg, New York, pp 581–640
2. Messenger JB (1991) Photoreception and vision in molluscs. In: Cronly-Dillon JR, Gregory RL (eds) Evolution of the eye and visual system. MacMillian, London, pp 364–397
3. Katagiri Y, Katagiri N, Fujimoto K (1985) Morphological and electrophysiological studies of a multiple photoreceptive system in a marine gastropod, *Onchidium*. Neurosci Res Suppl 2:S1–S15
4. Katagiri N, Hama K, Katagiri Y, et al. (1990) High-voltage electron microscopic study on the axon of the dermal photoreceptor cell in the dorsal mantle of *Onchidium verruculatum*. J Electron Microsc (Tokyo) 39:363–371
5. Hisano N, Tateda H, Kuwabara M (1972) Photosensitive neurones in the marine pulmonate mollusc *Onchidium verruculatum*. J Exp Biol 57:651–660
6. Stantschinsky W (1908) Über den Bau der Rückenaugen und die Histologie der Rückenregion der Oncidien. Z Wiss Zool 90:137–180
7. Eakin RM (1972) Structure of invertebrate photoreceptors. In: Dartnall HJA (ed) Handbook of sensory physiology, vol. 7/1: Photochemistry of vision. Springer-Verlag, Berlin, Heidelberg, New York, pp 625–684
8. Katagiri N, Katagiri Y, Shimatani Y, et al. (1995) Cell type and fine structure of the retina of *Onchidium* stalk-eye. J Electron Microsc (Tokyo) 44:219–230
9. Katagiri N (1984) Cytoplasmic characteristics of three different rhabdomeric photoreceptor cells in a marine gastropod, *Onchidium verruculatum*. J Electron Microsc (Tokyo) 33:142–150
10. Röhlich P, Aros B, Viragh Sz (1970) Fine structure of photoreceptor cells in the earthworm *Lumbricus terrestris*. Z Zellforsch Mikrosk Anat 104:345–357
11. Pentreath VW (1995) Functions of invertebrate glia. In: Ali MA (ed) Nervous systems of invertebrates. Plenum Press, New York, pp 61–103

The Cystein-Rich Domains of RDGA, a *Drosophila* Eye-Specific Diacylglycerol Kinase, Are Required for Protein Localization and Enzyme Activity

EMIKO SUZUKI[1], ICHIRO MASAI[2], HIROKO INOUE[3], and TAKESHI AWASAKI[4]

Key words. Retinal degeneration, Diacylglycerol kinase, *Drosophila*

Retinal DeGeneration A (RDGA) is a *Drosophila* eye-specific diacylglycerol kinase (DGK), whose genetic mutation blocks photoreceptor potentials and causes severe retinal degeneration. *Drosophila* phototransduction is mediated by the phosphoinositide (PI) signaling, initiated by the G-protein coupled phospholipase C (NORPA). Within this cascade, RDGA mediates an initial step for the regeneration of PIs by the production of phosphatidic acid, and attenuates the intracellular level of DG that regulates the activity of the light-sensitive membrane channels (TRP channels). In a photoreceptor cell, RDGA protein localizes to the restricted subcellular region called the subrhabdomeric (SR) region that is adjacent to the photoreceptive organelle, the rhabdomere. This region contains a specialized network of smooth endoplasmic reticulum, subrhabdomeric cistern (SRC), and RDGA is thought to function in association with the membranes of SRC. Thus, the localization of RDGA to the SR region appears to be important for its function. We have asked which domain of RDGA protein is necessary for its subcellular localization and function. The structure of RDGA protein predicted from the gene contains tandemly arranged cystein-rich domains (CRDs) in the N-terminal region, a catalytic domain in the middle region, and four ankyrin-like repeats at the C-terminus. The rescue experiments by germ-line transformation on *rdgA* null mutants expressing RDGA proteins that are deleted in these domains showed that the CRDs are necessary for the intracellular localization and function of RDGA. The RDGA protein without these domains does not rescue the degeneration phenotype, shows no recovery of DGK activity, and mislocalizes in the cytoplasm of photoreceptors. This is, to our knowledge, the first demonstration of the function of CRDs of DGKs in vivo.

[1] Division of Structural Biology, Institute of Medical Science, University of Tokyo, and CREST JST, 4-6-1 Shirokanedai, Minato-ku, Tokyo 108-8639, Japan
[2] Masai Initiative Research Unit, RIKEN, 2-1 Hirosawa, Wako, Saitama 351-0198, Japan
[3] School of Human Science, Waseda University, 2-579-15 Mikajima, Tokorozawa, Saitama 359-1192, Japan
[4] National Institute of Basic Biology, Myodaiji, Okazaki, Aichi 444-8585, Japan

Polymorphic Variations in Long- and Middle-Wavelength-Sensitive Opsin Gene Loci in Crab-Eating Monkeys

AKISHI ONISHI[1,2], KENICHI TERAO[3], OSAMU TAKENAKA[4],
AKICHIKA MIKAMI[5], SHUNJI GOTO[6], HIROO IMAI[1,2],
YOSHINORI SHICHIDA[1,2], and SATOSHI KOIKE[3]

Key words. Old World primate, Color vision deficiency, Polymorphic variation

Old World primates have trichromatic color vision as a result of three types of cone opsin, each of which is maximally sensitive to either short (S), middle (M), or long (L) wavelengths. Most color vision defects in humans are L/M abnormal that are caused by unequal crossing over between L and M visual pigment genes. We previously reported that only three crab-eating monkeys of the 2788 macaque monkeys are protanopic (L pigment absent) and that the frequency of dichromats in this species is lower than that in humans [1]. To understand the mechanism leading to the difference in L and M genes between humans and crab-eating monkeys more profoundly, we investigated different genotype variants of the L and M genes in the crab-eating monkeys using 758 available DNA samples of this species. We first identified polymorphisms in the copy number of L and M genes. In humans, about 60% of males have a gene array consisting of a single L gene and more than one M opsin gene, but only 5% exhibit this genotype in crab-eating monkeys. We next investigated anomalous trichromats by detecting amino acid substitutions on exon 5 that are responsible for L/M spectral tuning (sites Y277F and T285A). As a result, there were no anomalous trichromats, while about 6% of humans have such color vision defects. In addition, we estimated amino acid substitution at S180A in exon 3, because the substitution causes spectral shift but does not result in color vision defects. In humans, about 40% of males have such a substitution but only 1% of crab-eating monkeys do. The percentages of each variant were quite different between the two species, and the variants were found in locally restricted areas in crab-eating monkeys.

---

[1] Department of Biophysics, Graduate School of Science, Kyoto University, Kyoto 606-8502, Japan
[2] Core Research for Evolutional Science and Technology (CREST), Japan Science and Technology Corporation, Kyoto 606-8502, Japan
[3] Department of Microbiology and Immunology, Tokyo Metropolitan Institute for Neuroscience (TMIN), 2-6 Musashidai, Fuchu, Tokyo 183-8526, Japan
[4] Department of Cellular and Molecular Biology, [5] Department of Behavioral and Brain Sciences, and [6] Center for Human Evolutionary Modeling Research, Primate Research Institute, Kanrin, Inuyama, Aichi 484-8506, Japan

Reference

1. Onishi A, Koike S, Ida M, et al. (1999) Dichromatism in macaque monkeys. Nature 402:139–140

S- and M-Cone Electroretinograms in rd7 Mice with *NR2E3* Gene Mutation

Shinji Ueno, Mineo Kondo, Asahiko Takahashi, and Yozo Miyake

Key words. Electroretinogram, rd7 mice, S-cone

Purpose

Earlier studies have shown that the alterations of the visual function in rd7 mice result from a deletion in the photoreceptor cell-specific nuclear receptor gene, *Nr2e3* [1]. Mutations of this gene have been associated with the enhanced S-cone syndrome (ESCS) in humans [2]. Immunohistological studies also showed that the rd7 retina has an increased number of S-cone cells [3]. The purpose of this study was to examine the physiological properties of the S- and M-cones in rd7 mice electrophysiologically.

Methods

Scotopic and photopic intensity-response curves were obtained from electroretinograms (ERGs) recorded from 8–10 weeks rd7 ($n = 6$) and control mice ($n = 6$). S- and M-cone ERGs were elicited by photostrobe flashes with 410- and 530-nm narrow-band interference filters (half-bandwidth, 10 nm) on a rod-suppressing white background.

Results

Rod responses were recordable in the rd7 mice, although the a-waves were reduced to about one-half of the control, and the b-waves to 80% of the control. The amplitudes of the cone responses for rd7 did not differ from those for control mice. The amplitudes of both S- and M-cone ERGs of the rd7 mice did not differ from those of the control mice.

Department of Ophthalmology, Nagoya University School of Medicine, 65 Tsurumai-cho, Showa-ku, Nagoya 466-8550, Japan

Conclusions

These results indicate that rd7 mice do not demonstrate a prominent "enhanced S-cone" function as in human ESCS.

References

1. Akhmedov NB, Piriev NI, Chang B, et al. (2000) A deletion in a photoreceptor-specific nuclear receptor mRNA causes retinal degeneration in the rd7 mouse. Proc Natl Acad Sci USA 97:5551–5556
2. Haider NB, Jacobson SG, Cideciyan AV, et al. (2000) Mutation of a nuclear receptor gene, *NR2E3*, causes enhanced S cone syndrome, a disorder of retinal cell fate. Nat Genet 24:127–131
3. Haider NB, Naggert JK, Nishina PM (2001) Excess cone cell proliferation due to lack of a functional *NR2E3* causes retinal dysplasia and degeneration in rd7/rd7 mice. Hum Mol Genet 10:1619–1626

Activation of Group III Metabotropic Glutamate Receptor Suppresses the Membrane Currents in Newt Photoreceptors

Nobutake Hosoi, Jeongchul Hong, and Masao Tachibana

Key words. Metabotropic glutamate receptor, Photoreceptor, Hyperpolarization-activated current

It is well established that ON-type and OFF-type bipolar cells receive glutamatergic inputs from photoreceptors via a subtype of group III metabotropic glutamate receptors (mGluRs), mGluR6, and ionotropic glutamate receptors (non-NMDA receptors), respectively. However, it has been reported recently that a group III mGluR selective agonist L-2-amino-4-phosphonobutyrate (L-AP4) not only suppresses the ON responses but also affects the OFF responses of ganglion cells, and that the group III mGluR (mGluR8) is expressed at the terminal of rat photoreceptors. In the newt retina, we confirmed that L-AP4 not only abolished the ON responses but also reduced the OFF responses of ON-OFF ganglion cells. Furthermore, L-AP4 reduced the amplitude of the light-evoked currents but not the glutamate-induced current in OFF bipolar cells. These data suggest that activation of the group III mGluR by L-AP4 may decrease the amount of glutamate release from photoreceptors. In the present study we investigated how L-AP4 affects the glutamate release from photoreceptors. Using a newt retinal slice preparation, we measured the membrane currents of photoreceptors, namely, the calcium current (I_{Ca}) and the hyperpolarization-activated current under the whole-cell voltage clamp. Application of L-AP4 reduced the L-type I_{Ca} in cones but not in rods. On the other hand, the current activated by membrane hyperpolarization was suppressed by L-AP4 in both cones and rods. This current was identified as I_h since the current was blocked by external application of Cs^+ or ZD7288, a specific blocker of I_h, but not by Ba^{2+}. The effects of L-AP4 on I_h, were ascribed to the slowing of the activation kinetics and to the negative shift of the activation curve along the voltage axis. These results suggest that activation of the group III mGluR may reduce the glutamate release by inhibiting I_{Ca} and/or I_h in photoreceptors.

Department of Psychology, Graduate School of Humanities and Sociology, The University of Tokyo, 7-3-1 Hongo, Bunkyo-ku, Tokyo 113-0033, Japan

Quantitative Analysis of Signal Transduction in the Olfactory Receptor Cell

Hiroko Takeuchi and Takashi Kurahashi

Key words. Caged cyclic AMP, Signal transduction, CNG channel

We investigated the effect of cyclic nucleotide monophosphate (cNMP) on the cytoplasm of cilia in olfactory receptor cells. In this experiment, we applied caged cNMP to isolated living olfactory receptor cells with the whole-cell patch-clamp method. Photolysis of caged compounds was controlled quantitatively by UV light stimulation locally applied to the cilia. Light illumination induced an inward current in all tested cells, which were loaded with either 1 mM caged cyclic adenosine monophosphate (cAMP) or 1 mM caged cyclic guanosine monophosphate. The amplitude of the light-induced current was dependent on both light intensity and duration. The intensity- and duration-response relation were well fitted by the Hill equation with high cooperativity (Hill coefficient, approximately 4–5), supporting the notion that a Cl current added onto a cNMP-induced current makes a nonlinear booting of the signal transduction. To confirm that idea, we examined Hill's fitting for the developing phase of the current at +100 mV and −50 mV. The developing phase became more slowly at +100 mV, and the time course was fitted by a smaller Hill coefficient. Furthermore, we compared the light- and odorant-induced response to estimate the activation time course of adenylyl cyclase. Adenylyl cyclase was activated about 260 ms after the onset of the odorant stimulation. When long steps (light and odorant) were applied to the cells, the odorant-induced current showed a stronger and more remarkable decay than did the light-induced response. This observation suggests that the molecular system regulating desensitization locates upstream of the cAMP production site in addition to the direct modulation of the CNG channel.

Physiological Laboratory Department of Frontier Biosciences, Osaka University, 1-3 Machikaneyama, Toyonaka, Osaka 660-8531, Japan

RPE65-Related Proteins Constitute a Retinoid- and Carotenoid-Metabolizing Protein Family Throughout the Animal and Plant Kingdom

Hiroshi Sagara, Emiko Suzuki, and Kazushige Hirosawa

Key words. RPE65, Retinoid, Carotenoid, *Drosophila*, DRPE65

Introduction

Retinal pigment epithelial (RPE) cells play crucial roles in the maintenance of visual functions. The production and regeneration of 11-*cis* retinal, which serves as the chromophore of the visual pigment rhodopsin, is one of the most important roles of the RPE cells. To date, several genes that encode the enzymes that may participate in the metabolism of retinoids in RPE cells have been identified [1–3]. However, some of the key enzymes presumed to exist in the RPE cell such as isomerohydrolase have not yet been identified. RPE65 is one of the candidate proteins that may be included in retinoid metabolism in RPE cells.

RPE65 was first described in the chick retina by Sagara and Hirosawa [4] as a 63-kDa protein abundantly and specifically expressed in RPE cells. We and others have identified the RPE65 genes in several vertebrate species. Analysis of the predicted amino acid sequences has shown that RPE65s are highly conserved from fish to primates (more than 80% identical). Mutations in the human RPE65 gene have been shown to result in severe retinal dystrophy [5], indicating that this protein is important for vertebrate retinal function. RPE65-deficient mice show abnormal accumulation of all-*trans* retinyl ester in RPEs [6]. This suggests the involvement of RPE65 in the retinoid cycle in the eye. However, the precise function of RPE65 is elusive.

In this chapter, we describe our studies on RPE65s of various vertebrate species, and on the identification of the *Drosophila* gene (*drpe65*) that is highly homologous to vertebrate RPE65s. Our data and analysis of the genome database strongly suggest that RPE65-related protein constitutes a retinoid- and carotenoid-metabolizing protein family throughout the animal and plant kingdom.

Division of Fine Morphology, Department of Basic Medical Sciences, The Institute of Medical Science, The University of Tokyo, 4-6-1 Shirokanedai, Minato-ku, Tokyo 108-8639, Japan

Materials and Methods

Three monoclonal antibodies raised against chick [4] and bovine RPE cells were used to study the variability of the antigens in the RPE cells of a wide variety of animals from fish to mammalian classes. The antibodies were also used for immunoelectron microscopy to determine the intracellular localization of the antigen within the chick RPE cells. To identify the cDNA sequence of the antigen, we screened a chick RPE cell cDNA library with these antibodies. The zebrafish homologue of the RPE65 gene was identified by screening a zebrafish eye cDNA library using a chick cDNA fragment as a probe.

cDNA clones of the *Drosophila* homologue of vertebrate RPE65s were obtained by the degenerate polymerase chain reaction (PCR) method followed by screening the *Drosophila* head cDNA library with the obtained PCR fragments.

Results and Discussion

Vertebrate RPE65s

The retina of animals from a wide variety of classes were examined for their reactivities to the monoclonal antibodies raised against chick and bovine RPE cells (Table 1). The antibody raised against bovine RPE cells (named Y3H) recognized the antigen in the RPE cell cytoplasm from almost all of the animals examined, showing that the antigenic epitope was highly conserved. The only vertebrate eye not stained by Y3H was that of lamprey. S5D8, an antibody raised against chick RPE cells, also recognized the antigen in RPE cells from many vertebrate animals. S5D8 failed to stain RPE cells from newt, carp, and lamprey, all of which are known to use vitamin A$_2$ (3–4-dehydroretinal) as the visual chromophore. Although classified into the same class as carp, RPE cells from tai fish and zebrafish, known to use vitamin A$_1$ (retinal) like other terrestrial animals, were stained by S5D8, suggesting the vitamin A$_1$-related functions of the epitope recognized by S5D8. Another antibody raised against chick RPE cells (S5H8) stained the RPE cells of mammals, birds, and reptiles but failed to stain amphibian and fish RPE cells, suggesting that the epitope recognized by S5H8 was newly evolved. At this time, the epitopes recognized by these monoclonal antibodies in the antigenic molecules is not known. Identifying these epitopes may help in understanding the function of the antigenic molecules.

When examined by immunoelectron microscopy [4], the antigens recognized by S5D8 and S5H8 were localized on the membrane of the smooth endoplasmic reticulum where the vitamin A-metabolizing reactivities reside. Immunoreaction products were also observed on the membrane of the rough endoplasmic reticulum but not on the nuclear membrane or on the plasma membrane.

To determine the molecular nature of the antigen recognized by these antibodies, we tried to identify the genes that encode the antigenic proteins. A chick RPE cell cDNA library was screened with these antibodies and eight cDNA clones were obtained. Nucleic acid sequence analysis revealed that five of these clones were independent cDNA fragments coding the same protein. The nucleic acid sequence of the cDNA had high homology with the sequences of *RPE65s*, also known as RPE-specific

TABLE 1. Reactivities of three monoclonal antibodies against retinal pigment epithelial cells of various species of animals. + represents immuno-positive, − represents immuno-negative

	S5D8	S5H8	Y3H
Squirrel monkey	+	+	+
Cow	+	+	+
Rabbit	+	+	+
Mouse	+	+	+
Chick	+	+	+
Tortoise	+	+	+
Newt	−	−	+
Tai fish	+	−	+
Carp	−	−	+
Zebrafish	+	−	+
Lamprey	−	−	−

proteins. The amino acid sequence predicted from the nucleic acid sequence had about 90% identity with the amino acid sequences of RPE65s from bovine, human, and rat. From the high homology of the amino acid sequence and the specific expression of the protein in RPE, we concluded that the obtained cDNA clone was the chick homologue of *RPE65*. The nucleic acid sequence has been submitted to the DDBJ/GenBank/EMBL Data Bank with accession number AB017594.

To confirm that these antibodies really recognize the protein encoded by the obtained cDNA, we transfected golden hamster embryonic lung cells with the expression vector containing a full-length open reading frame of the cDNA and examined them by immunofluorescence microscopy. All three antibodies stained the transfected cells, showing that the protein encoded by this cDNA, RPE65, was really the antigen recognized by these three antibodies.

Recently, beta-carotene dioxygenase genes of several vertebrates were identified. The amino acid sequence of the beta-carotene dioxygenase had high homology (about 70% identity) with that of RPE65 of the same species. Beta-carotene dioxygenase is the enzyme that oxidatively cleaves the carbon double bond at the 15,15′ position of beta-carotene to yield two molecules of retinal. Thus, there are two types of RPE65-related proteins in vertebrates.

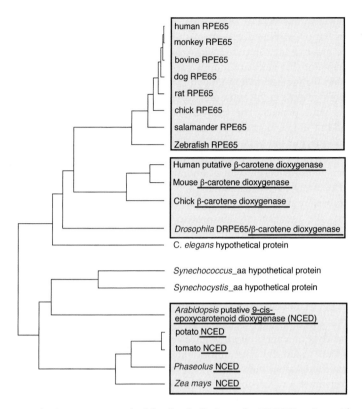

FIG. 1. Protein databases were searched for the similarity to the DRPE65 amino acid sequence, and evolutional distances were calculated using sequence analyzing program GENETYX-MAC. There are two groups known as the enzymes involved in carotenoid metabolism (*underlined*)

When crayfish eyes were examined for antigenicity against these antibodies, we found immunoreactivity around the screening pigments in the photoreceptor cells. The highly conserved nature of the amino acid sequence of vertebrate RPE65s and the RPE65-like immunoreactivity in the crayfish eye encouraged us to explore the invertebrate homologue of RPE65.

Invertebrate Homologue of RPE65

To investigate the invertebrate homologue of RPE65, we chose *Drosophila melanogaster* as the experimental animal because of their accumulated data as an model animal in genetic studies and the relative ease of later functional analysis.

By degenerate PCR and following a screen of the *Drosophila* head cDNA library, we isolated two clones that included the same gene with homology to RPE65s of vertebrates. The predicted amino acid sequence of this gene had about 40% identity with those of the vertebrate RPE65s. Northern blot analysis showed that the transcripts were expressed mainly in the head; a detectable amount was not found in the body.

Reverse transcriptase-PCR analysis using eyes absent (*eya*), a mutant lacking compound eyes but preserving ocelli, showed drastic reduction of the expression of this gene in the *eya* mutant compared with wild-type animals. This result indicates that the transcripts of this gene are mainly expressed in the eye. The high homology of this gene with the vertebrate *RPE65s* and the expression of this gene in the eye suggest that the obtained gene is the *Drosophila* orthologue of the vertebrate *RPE65*. Thus, we named this gene *drpe65*. The nucleic acid sequence has been submitted to the DDBJ/GenBank/EMBL Data Bank with accession number AB041507. A recent study aimed at identifying the beta-carotene dioxygenase gene in *Drosophila* showed that the same protein as DRPE65 had the beta-carotene dioxygenase activity [7]. In *Drosophila melanogaster*, the whole-genome sequence has been clarified and no other type of RPE65-related protein other than DRPE65/beta-carotene dioxygenase was found. The preferential expression of the DRPE65/beta-carotene dioxygenase in the eye suggests the function of this protein is related to vision. In fact, *ninaB*, the *Drosophila* sight-defective mutant, was reported to have a mutation in the beta-carotene dioxygenase gene [8].

In addition to the vertebrate RPE65s and beta-carotene dioxygenases, the amino acid sequence alignment analysis revealed several regions conserved among drpe65 and plant 9-*cis* epoxycarotenoid dioxygenases. 9-*cis* epoxycarotenoid dioxygenase is a plant enzyme also related to carotenoid metabolism. Putative proteins with homology to RPE65 were also predicted from the genome sequences of *Caenorhabditis elegans* and Cyanobacteria. Homologous regions and their arrangement within the molecules were much the same among these proteins, suggesting that these regions are motifs with important function(s). These observations indicate that RPE65-related proteins constitute a carotenoid- and retinoid-metabolizing protein family throughout the animal and plant kingdoms (Fig. 1).

References

1. Simon A, Hellman U, Wernstedt C, et al. (1995) The retinal pigment epithelial-specific 11-*cis*-retinol dehydrogenase belongs to the family of short chain alcohol dehydrogenases. J Biol Chem 270:1107–1112
2. Haeseleer F, Huang J, Lebioda L, et al. (1998) Molecular characterization of a novel short-chain dehydrogenase/reductase that reduces all-*trans*-retinal. J Biol Chem 273:21790–21799
3. Ruin A, Winston A, Lim Y-H, et al. (1999) Molecular and biochemical characterization of lecithin retinol acyltransferase. J Biol Chem 274:3834–3841
4. Sagara H, Hirosawa K (1991) Monoclonal antibodies which recognize endoplasmic reticulum in the retinal pigment epithelium. Exp Eye Res 53:765–771
5. Redmond TM, Hamel CP (2000) Genetic analysis of RPE65: from human disease to mouse model. Methods Enzymol 316:705–724
6. Redmond TM, Yu S, Lee E, et al. (1998) Rpe65 is necessary for production of 11-*cis*-vitamin A in the retinal visual cycle. Nat Genetics 20:344–351
7. Lintig J, Vogt K (2000) Filling the gap in vitamin A research. J Cell Biol 257:11915–11920
8. Lintig J, Dreher A, Kiefer C, et al. (2001) Analysis of the blind *Drosophila* mutant ninaB identifies the gene encoding the key enzyme for vitamin A formation in vivo. Proc Natl Acad Sci USA 98:1130–1135

Simulation Analysis of Retinal Horizontal Cells

Toshihiro Aoyama[1], Yoshimi Kamiyama[2], and Shiro Usui[1,3]

Key words. Horizontal cells, Simulation, Mathematical model, Ionic current, Membrane property

Introduction

Retinal horizontal cells (HCs) are second-order neurons, which play a fundamental role in forming center-surround receptive fields. Ionic current is an important element for deciding on electrical membrane characteristics. Ionic channels on HCs have been investigated by applying the patch-clamp method and pharmacology techniques. However, it is difficult to elucidate the function of each ionic current for electrical behaviors of the neurons by only using these techniques. The physiological engineering approach is highly effective for understanding ionic current mechanisms. Mathematical models based on physiological knowledge, such as ionic current mechanisms and intracellular information processing systems, are used to understand complicated electrical behavior in neurons. This paper summarizes the studies on analyzing the membrane properties of HCs.

Ionic Currents of Horizontal Cells

Ionic currents of HCs have been measured in many species (summarized in Table 1). In most animals, five types of voltage-dependent ionic currents have been mainly found in the cell body: a sodium current (I_{Na}), an L-type calcium current (I_{CaL}), a delayed rectifying potassium current (I_{Kv}), a transient outward potassium current (I_A), and an anomalous rectifying potassium current (I_{Ka}). The properties that depend on the voltage of these currents are similar among animals, but the maximum conductances are different. Another difference is the dynamics of I_{CaL}. I_{CaL} is inactivated by

[1] Brain Science Institute, RIKEN, 2-1 Hirosawa, Wako, Saitama 351-0198, Japan
[2] Faculty of Information Science and Technology, Aichi Prefectural University, Nagakute, Aichi 480-1198, Japan
[3] Department of Information and Computer Sciences, Toyohashi University of Technology, 1-1 Tenpaku, Toyohashi, Aichi 441-8580, Japan

TABLE 1. Ionic currents of horizontal cells

Abbreviation	Current name	Species (references)
I_{Na}	Sodium	Goldfish [1], catfish [2], cat [3], rabbit [4], human [5]
I_{Ca_L}	L-type calcium	Goldfish [1, 6], catfish [2], white bass [7], cat [3] rabbit [4], human [5]
I_{Ca_T}	T-type calcium	White bass [7], rabbit [4]
I_{Kv}	Delayed rectifying potassium	Goldfish [6, 1], white bass [8], cat [3], rabbit [4], human [5]
I_A	Transient outward potassium	Goldfish [6, 1], white bass [8], cat [3], rabbit [4], human [5]
I_{Ka}	Anomalous rectifying potassium	Goldfish [6, 1], human [5], cat [3], rabbit [4]
I_{Kc}	Calcium-dependent potassium	Rabbit [4]

the accumulation of intracellular calcium ions in goldfish HCs [6]; however, this inactivation has not been observed in other animals.

Horizontal Cell Model

The ionic current models of HCs were developed for goldfish [9–11] and rabbit [12]. Five types of ionic currents in retinal HCs were described by Hodgkin-Huxley-type equations based on voltage-clamp measurements of goldfish [6] and rabbit [4, R. Blanco, P. de la Villa, C. Vaquero, unpublished]. The HC models were developed by integrating these five types of ionic current models into a parallel conductance model. The HC models of goldfish [10, 11] also include the intracellular calcium system to reproduce the inactivation of I_{Ca}. The HC models are capable of accurately reproducing electrophysiological experiments: voltage- and current-clamp experiments, dose-response to glutamate, and so on [9, 11, 12].

Simulation Analysis

The membrane potential response of neurons results from the activation of ionic currents. The model based on ionic current mechanisms is useful for elucidating the complicated electrical behavior of neurons. The following results are a few examples of simulation analysis for retinal HC responses.

Contributions of Ionic Current Nonlinearities to the V-I Relationship [11]

The dissociated goldfish HCs showed a nonlinear steady-state voltage-current (V-I) relationship in standard solution [13]. Figure 1A shows that the model that includes all ionic currents reproduces the V-I relationship well in an electrophysiological experiment [13].

Figure 1B–D indicates how an ionic current affects the V-I relationship of HCs. If the cell is constructed as a simple membrane model consisting only of leakage current (I_l), the V-I relationship is linear (solid line in each figure). Each ionic current non-

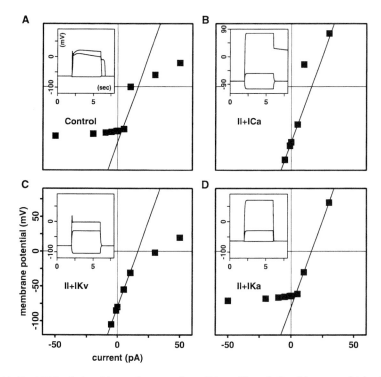

FIG. 1A–D. A V–I relationships under control conditions. The relationships were obtained from the voltage responses (*inset*). The membrane potentials were measured at the termination of the current pulses. I_l, leakage current, I_{Ca}, calcium current; I_{Kv}, transient outward potassium current

linearizes the linear V-I relationship, depending on its voltage dependency. In membrane potential depolarized over 0 mV, the conductance is larger because of activation of I_{Kv} (C). The increment of conductance under –60 mV results from the activation of I_{Ka} (D). The transient currents, I_{Na} and I_A, did not affect the steady-state V-I relationship (not shown). I_{Ca} causes a hyperpolarizing jump [13] (B). The result also shows that the steep slope that was observed in Ca-free solution [13] is due to I_l conductance.

Repetitive Action Potential [12]

A repetitive action potential was observed in dissociated rabbit A-type HCs [14]. In response to a large-amplitude current injection, an action potential or repetitive action potentials were observed in 7 of 17 HCs. The long-lasting depolarization was observed in the rest of the cells. The underlying mechanism of spike generation was not clarified at that time. Figure 2 shows the voltage responses to current injection in the rabbit HC model with a large conductance of I_{Kv}. When a current pulse with amplitude over 25 pA was applied, an action potential or repetitive action potentials were produced by the model. These responses are highly similar to those measured by

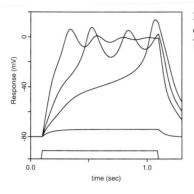

FIG. 2. Current clamp simulations under a large conductance of I_{Kv}. The current was 10, 25, 40, and 70 pA (1 s)

experiment [14]. Simulation results show that the delayed rectifying potassium current (I_{Kv}) and the calcium current (I_{CaL}) are responsible for the repetitive action potentials [12].

Conclusion

The dissociated HCs show complicated responses to current injection. The simulation results revealed which ionic current induced the sigmoidal membrane characteristic and the repetitive calcium spike. It will also be important to elucidate how the ionic current mechanisms affects retinal visual information processing. The model based on physiological studies will be a useful tool for studying processing.

References

1. Yagi T, Kaneko A (1988) The axon terminal of goldfish retina horizontal cells: a low membrane conductance measured in solitary preparations and its implication to the signal conduction from the soma. J Neurophysiol 59:482–494
2. Shingai R, Christensen B (1983) Sodium and calcium current measured in isolated catfish horizontal cells under voltage clamp. Neuroscience 10:893–897
3. Ueda Y, Kaneko A, Kaneda M (1992) Voltage-dependent ionic currents in solitary horizontal cells isolated from cat retina. J Neurophysiol 68:1143–1150
4. Löhrke S, Hofmann H (1994) Voltage-gated currents of rabbit A- and B-type horizontal cells in retinal monolayer cultures. Vis Neurosci 11:369–378
5. Picaud S, Hicks D, Forster V, Sabel J, Dreyfus H (1998) Adult human retinal neurons in culture: physiology of horizontal cells. Invest Ophthalmol Vis Sci 39:2637–2648
6. Tachibana M (1983) Ionic currents of solitary horizontal cells isolated from goldfish retina. J Physiol 345:329–351
7. Sullivan J, Lasater E (1992) Sustained and transient calcium currents in horizontal cells of the white bass retina. J Gen Physiol 99:85–107
8. Sullivan J, Lasater E (1990) Sustained and transient potassium currents of cultured horizontal cells isolated from white bass retina. J Neurophysiol 64:1758–1766
9. Usui S, Kamiyama Y, Ishii H, Ikeno H (1996) Reconstruction of retinal horizontal cell responses by the ionic current model. Vision Res 36:1711–1719
10. Hayashida Y, Yagi T (2002) On the interaction between voltage-gated conductances and Ca^{2+} regulation mechanisms in retinal horizontal cells. J Neurophysiol 87:172–182

11. Aoyama T (2002) Physiological engineering study of retinal horizontal cell response based on the nonlinear membrane properties. PhD thesis, Toyohashi University of Technology
12. Aoyama T, Kamiyama Y, Usui S, et al. (2000) Ionic current model of rabbit retinal horizontal cell. Neurosci Res 37:141–151
13. Tachibana M (1981) Membrane properties of solitary horizontal cells isolated from goldfish retina. J Physiol 321:141–161
14. Blanco R, Vaquero C, de la Villa P (1996) Action potentials in axonless horizontal cells isolated from the rabbit retina. Neurosci Lett 203:57–60

External Proton Mediates the Feedback from Horizontal Cells to Cones in the Newt Retina

Hajime Hirasawa and Akimichi Kaneko

Key words. Horizontal cell, Cone, Feedback

The center-surround organization of the receptive fields is one of the most important properties of retinal neurons. Cone photoreceptors show concentric receptive field; they are hyperpolarized by spot light and depolarized by annulus light. It is widely accepted that the negative feedback from horizontal cells (HCs) generates the surround response of cones. Although accumulating evidence suggests that γ-aminobutyric acid (GABA) mediates the feedback from HC to cones, there are still a number of reports criticizing the GABA hypothesis of HC-cone feedback. In the goldfish, Verweij et al. [1] have reported that the feedback was detectable in the presence of GABA antagonist, and the surround illumination shifted the activation voltage of the cone calcium current (I_{Ca}) to be more negative. In the present study, we reexamined the mechanism of feedback to find an answer to this controversy.

Cone photoreceptors were recorded by the whole-cell patch-clamp technique in newt retinal slices, in the presence of GABA antagonist (100 μM picrotoxin). Surround illumination shifted the activation voltage of I_{Ca} to a more negative voltage range. The shift of the activation voltage of I_{Ca} was not merely a parallel shift of the current-voltage (I–V) relation along the voltage axis, because the surround illumination increased I_{Ca} even at voltages more positive than the I_{Ca} peak. Focal application of basic solution (pH 9.0) to the cone synaptic terminal layer increased I_{Ca} in a similar manner as the surround illumination. These observations suggest a possibility that surround illumination lowered the proton concentration in the cone synaptic terminal region. Addition of external 10 mM hydroxyethylpiperazine ethanesulfonic acid, which increased the pH buffering capacity, suppressed cone surround response without affecting either the light response of cones or the I_{Ca} recorded in the presence of surround illumination. Both 20 μM kainate and 10 μM 6-cyano-7-nitroquinoxaline (CNQX), an agonist and antagonist of α-amino-3-hydroxy-5-methyl-4-isoxazole proprionic acid (AMPA) receptors of HCs, suppressed surround responses of cones. However, while CNQX (HC hyperpolarized) augmented cone Ica, kainate (HC depolarized) suppressed cone I_{Ca}.

Department of Physiology, School of Medicine, Keio University, 35 Shinanomachi, Shinjuku-ku, Tokyo 160-8582, Japan

From these results, we propose a hypothesis that the proton is the mediator of feedback from HCs to cones. According to this hypothesis, following sequence of events is expected. HCs are depolarized in the dark, the synaptic cleft is maintained in an acidified condition, cone I_{Ca} is suppressed, and the transmitter release from cones is low. Upon surround illumination, HCs are hyperpolarized, making the synaptic cleft more basic, which increases I_{Ca} and transmitter release from the cone terminal. The mechanisms that control the proton concentration in relation to the HC membrane voltage needs to be elucidated.

Reference

1. Verweij J, Kamermans M, Spekreijse H (1996) Horizontal cells feed back to cones by shifting the cone calcium-current activation range. Vision Res 36:3943–3953

A Metabotropic Glutamate Receptor Mediates Synaptic Transmission from Blue-Sensitive Cones to ON-Bipolar Cells in Carp Retina

Hiroshi Joho[1,2], Richard Shiells[3], and Masahiro Yamada[1]

Key words. mGluR, Blue-sensitive cones, ON-bipolar cells

Glutamate receptors mediate synaptic transmission from photoreceptors to bipolar cells in the vertebrate retina. L-2-amino-4-phosphonobutyrate (L-APB), a group III metabotropic glutamate receptor (mGluR) agonist, was used to identify mGluR6, which mediates rod synaptic input to ON-bipolar cells. A presynaptic mGluR has been similarly identified on red-sensitive cone synaptic terminals [1]. Whether a post-synaptic mGluR is involved in signaling from red-, green-, or blue-sensitive cones to ON-bipolar cells has not, however, been determined in the teleost retina. To characterize the postsynaptic glutamate receptor, electroretinogram (ERG) b-waves (a measure of ON-bipolar cell activity) and intracellular responses from cones to trichromatic stimuli (blue, 455 nm; green, 549 nm; and red, 694 nm) were recorded from light-adapted isolated carp retina.

Fifty micromolar L-APB selectively suppressed the b-wave elicited by blue light in the presence of $3\,\mu M$ 6-cyano-7-nitroquinoxaline-2,3-dione (CNQX), a non N-methyl-D-aspartate (NMDA) ionotropic glutamate receptor antagonist. This was used to suppress the activity of cone-driven OFF-bipolar and horizontal cells, which possess ionotropic glutamate receptors, and accordingly, this concentration was remarkably effective in reducing the d-wave (OFF-response) of the ERG. L-APB did not, however, suppress the intracellular photoresponses of cones in the presence of CNQX, indicating that L-APB suppressed the blue b-wave by a direct, postsynaptic action on the ON-bipolar cells making synaptic contact with blue-sensitive cones.

These results suggest that synaptic input from blue cones to ON-bipolar cells is mediated by an APB-sensitive mGluR in carp retina, similar to that mediating synaptic transmission from rods.

Reference

1. Hirasawa H, Shiells R, Yamada M (2002) A metabotropic glutamate receptor regulates transmitter release from cone presynaptic terminals in carp retinal slices. J Gen Physiol 119:55–68

[1] Tokyo Metropolitan Institute of Technology, 6-6 Asahigaoka, Hino, Tokyo 191-0065, Japan
[2] Yamanouchi Pharmaceutical Co. Ltd, 3-17-1 Hasune, Itahashi, Tokyo 174-8612, Japan
[3] Physiology Department, University College London, London, UK

Mathematical Model Study of Continuous Transmitter Release in the Synaptic Terminal of Goldfish Retinal Bipolar Cells

Tomokatsu Kawakita[1], Akito Ishihara[1], Shiro Usui[1], and Masao Tachibana[2]

Key words. Mathematical model, Transmitter release, Retinal bipolar cell

The giant synaptic terminals of Mb1 bipolar cells isolated from the goldfish retina are suitable for the physiological study of transmitter release mechanisms. It has been shown that there are multiple components of exocytosis and endocytosis with different kinetics and Ca^{2+} dependence. Thus, it is difficult to evaluate the contribution of each component to the transmitter release.

In this study, a mathematical model of transmitter release was constructed to explain the various aspects of transmitter release, such as the time course of capacitance changes and the Ca^{2+} dependence of transmitter release. We hypothesized that the release-ready pool (RRP) consisted of the fast and slow components (1200 and 4800 vesicles; [1, 2]) and the reserve pool (750,000 vesicles; [3]). The refilling of synaptic vesicles from the reserve pool to the slow component of RRP was assumed to be Ca^{2+}-independent. Exocytosis depended on the Ca^{2+} concentration [4] near the membrane, which was estimated by the model proposed by Ishinara et al. [5]. Endocytosis consisted of the fast and slow components [6]. Model parameters were estimated from the data obtained from physiological experiments.

This mathematical model could simulate the relationship between the duration of depolarization and the amount of capacitance changes as well as the kinetics of capacitance changes during depolarizing pulses. The model demonstrated that the refilling of synaptic vesicles of RRP from the reserve pool was essential to explain the saturation of capacitance jumps by a long depolarizing pulse. The simulation study indicates that the goldfish Mb1 bipolar cells are able to release transmitter continuously during a long depolarization without depletion of synaptic vesicles. The model proposed in this study may provide a fundamental and general model for analyzing the Ca^{2+}-dependent transmitter release mechanisms.

[1] Department of Information and Computer Sciences, Toyohashi University of Technology, 1-1 Hibarigaoka, Tempaku, Toyohashi, Aichi 441-8580, Japan
[2] Department of Psychology, Graduate School of Humanities and Sociology, The University of Tokyo, 7-3-1 Hongo, Bunkyo-ku, Tokyo 113-0033, Japan

References

1. Sakaba T, Tachibana M, Matsui K, et al. (1997) Two components of transmitter release in retinal bipor cells: exocytosis and mobilization of synaptic vesicles. Neurosci Res 27:357–370
2. von Gersdorff H, Sakaba T, Berglund K, et al. (1998) Submillisecond Kinetics of glutamate release from a sensory synapse. Neuron 21:1177–1188
3. Gomis A, Burrone J, Lagnado L (1999) Two actions of calcium regulate the supply of releasable vesiclesat the ribbon synapse of retinal bipolar cells. J Neurosci 19:6309–6317
4. Heidelberger R, Heinemann C, Neher E, et al. (1994) Calcium dependence of the rate of exocytosis in a synaptic terminal. Nature 371:513–515
5. Ishihara A, Kamiyama Y, Usui S (2000) What shapes light responses of the retinal on-type bipolar cell?—a simulation study. Society for Neuroscience 30th Annual Meeting Abstracts (NewOrleans) 26:664
6. Lagnado L, Gomis A, Job C (1996) Continuous vesicle cycling in the synaptic terminal of retinal bipolar cells. Neuron 17:957–967

The Role of Ionic Currents in Shaping Light Responses of Retinal Bipolar Cell

Akito Ishihara[1], Yoshimi Kamiyama[2], and Shiro Usui[1]

Key words. Retina, Mb1 bipolar cell, Mathematical model, Light responses

Introduction

Mb1 bipolar cell, which is a subclass of bipolar cells in the goldfish retina, dominantly receives inputs from rods [1], and responds to a spot of light with depolarization (ON-type) [2]. Recently, Mb1 bipolar cell has been shown to generate spontaneous calcium transients [3], and is capable of light-evoked calcium spike [4]. The spikes would be generated by ionic currents in their axon terminal (AT) [3, 4]. It is essential to analyze the mechanisms using a modeling simulation in order to understand how the ionic current of the cell plays a role in shaping the light responses. In this study, we analyzed the contribution of ionic mechanisms to light responses of Mb1 bipolar cell using a mathematical model based on ionic currents.

Methods

Since the Mb1 bipolar cell has a heterogeneous distribution of ionic mechanisms, we modeled the cell with two compartments, a soma and an AT (Fig. 1). The somatic compartment has been modeled in our previous studies [5, 6]. We developed the AT compartment based on their ionic currents. Each voltage-dependent current in the AT compartment was described using Hodgkin-Huxley-type equations [7]. Since there are Ca^{2+}-dependent currents in the AT, we also modeled an intracellular Ca^{2+} mechanism. The membrane potential of the soma (V_S) and the AT (V_T) is given by:

$$C_S \frac{dV_s}{dt} = I_{AX} - \left(I_{mg,S} + I_{h,S} + I_{Kv,S} + I_{l,S} \right)$$

$$C_T \frac{dV_T}{dt} = -\left(I_{AX} + I_{h,T} + I_{Ca,T} + I_{K(Ca),T} + I_{Cl(Ca),T} + I_{NC,T} + I_{ATP,T} + I_{l,T} + I_{Cl(Ca)} \right)$$

[1] Information and Computer Sciences, Toyohashi University of Technology, 1-1 Hibarigaoka, Tenpaku-cho, Toyohashi, Aichi 441-8580, Japan
[2] Faculty of Information Science and Technology, Aichi Prefectural University, Nagakute-cho, Aichi 480-1198, Japan

113

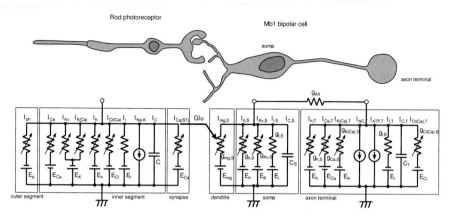

FIG. 1. The equivalent circuit of the Mb1 bipolar cell model with two compartments. When light stimulates the photoreceptor, the glutamate-induced current (I_{mg}) is activated by decreasing released glutamate from the photoreceptor model [8], and the Mb1 bipolar cell model depolarizes

where $I_{mg,s}$ is activated by the output of the rod photoreceptor model that has been described by Kamiyama et al. [8]. The parameters of the model were estimated from experimental data recorded from Mb1 bipolar cells [9–14].

Results

We examined the effect of the AT and each ionic current of the bipolar cell on light responses using the model (Fig. 1). Figure 2 illustrates simulated light responses to flashes of different intensity without the AT (Fig. 2A) and with the AT (Fig. 2B). With the AT, the amplitude decreased dramatically and shows a suppressed plateau level in high-intensity light. The results suggest that light responses of the bipolar cell are strongly affected by their AT.

In order to clarify which current affects the shape of the light responses, we analyzed individual ionic currents in the soma and the AT. We simulated the light responses with different values of the maximum conductance of ionic currents (25%–200% of estimated value), respectively (Fig. 3). When the ionic currents of the soma were changed (Fig. 3, top), light responses were affected only by I_{Kv} (Fig. 3B), and by the glutamate-induced current (Fig. 3A), which is directly mediated by the photoreceptors. I_h and I_l of both compartments had a slight effect on light responses (Fig. 3 C–E, I); the currents may not play much of a role in shaping the light responses. In the currents of the AT, I_{Ca} (Fig. 3F) and $I_{K(Ca)}$ (Fig. 3G) strongly affected the light responses. As I_{Ca} was decreased, the amplitude of the light responses was decreased (Fig. 3F). $I_{K(Ca)}$ had an effect on amplitude and waveform of light responses; the initial transient component was enhanced with decreased $I_{K(Ca)}$ (Fig. 3G). Decreasing $I_{Cl(Ca)}$ depolarized the plateau level of light responses (Fig. 3H). The results suggest that the light responses of the bipolar cell are shaped by not only somatic ionic current I_{Kv}, but also I_{Ca} and $I_{K(Ca)}$ of the AT.

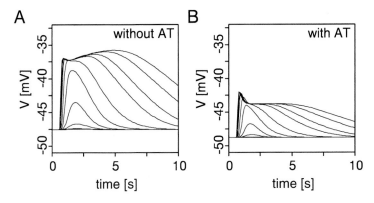

FIG. 2A,B. Simulated light responses of the bipolar cell with or without an axon terminal (*AT*). Light response to flashes with various intensities (duration 50 ms) was superimposed. A without AT ($g_{AX} = 0$), B with AT ($g_{AX} = 25$ [nS]).

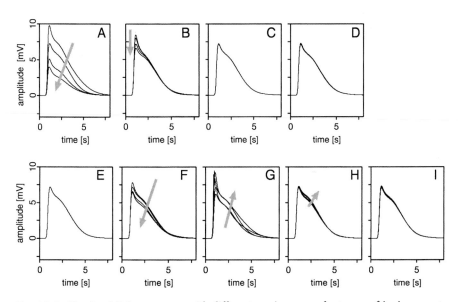

FIG. 3A–I. Simulated light responses with different maximum conductances of ionic current. *Arrow* indicates change in waveform with decreasing maximum conductance. **A** $g_{mg,s} = 2.0 - 15$ [nS]. **B** $g_{Kv,S} = 1.3 - 12$ [nS]. **C** $g_{h,S} = 0.25 - 2.0$ [nS]. **D** $g_{l,s} = 0.05 - 0.5$ [nS]. **E** $g_{h,T} = 0.25 - 2.0$ [nS]. **F** $g_{Ca,T} = 0.1 - 0.8$ [nS]. **G** $g_{K(Ca),T} = 3 - 25$ [nS]. **H** $g_{Cl(Ca),T} = 0.6 - 5$ [nS]. **I** $g_{l,T} = 0.1 - 0.6$ [nS]

Discussion

We developed an Mb1 bipolar cell model with two compartments, a soma and an AT. We examined the effect of the AT and each ionic current of the bipolar cell on light responses using the model. Analysis of ionic current components during the responses showed that the initial transient observed in a high-intensity light stimulus was formed by the delayed suppressive effect of Ca^{2+}-dependent currents. The results suggest that the light response is strongly affected by the AT and that I_{Ca} and $I_{K(Ca)}$ of the AT play a role in shaping the response.

References

1. Ishida AT, Stell WK, Lightfoot DO (1980) Rod and cone inputs to bipolar cells in goldfish retina. J Comp Neurol 1991:315–335
2. Saito T, Kondo H, Toyoda J (1979) Ionic mechanisms of two types of on-center bipolar cells in the carp retina. J Gen Physiol 73:73–90
3. Zenisek D, Matthews G (1998) Calcium action potentials in retinal bipolar neurons. Vis Neurosci 15:69–75
4. Protti DA, Flores-Herr N, Gersdorff H (2000) Light evokes Ca^{2+} spikes in the axon terminal of a retinal bipolar cell. Neuron 25:215–227
5. Ishihara A, Kamiyama Y, Usui S (1998) Analysis of light responses at the retinal bipolar cells based on ionic current model. In: Bower J (ed) Computational neuroscience: trends in research. Plenum Press, New York, 203–209
6. Usui S, Ishihara A, Kamiyama Y, et al. (1996) Ionic current model of bipolar cells in the lower vertebrate retina. Vision Res 36:4069–4076
7. Hodgkin AL, Huxley AF (1952) A quantitative description of membrane current and its application to conduction and excitation in nerve. J Physiol 117:500–544
8. Kamiyama Y, Ogura T, Usui S (1996) Ionic current model of the vertebrate rod photoreceptor. Vision Res 36:4059–4068
9. Kobayashi K, Tachibana M (1995) Ca^{2+} regulation in the presynaptic terminals of goldfish retinal bipolar cells. J Physiol 483:79–94
10. Tachibana M, Okada T, Arimura T, et al. (1993) Dihydropyridine-sensitive calcium current mediates neurotransmitter release from bipolar cells of the goldfish retina. J Neurosci 13:2898–2909
11. Okada T, Horiguchi H, Tachibana M (1995) Ca^{2+}-dependent Cl^- current at the presynaptic terminals of goldfish retinal bipolar cells. Neurosci Res 23:297–303
12. Sakaba T, Ishikane H, Tachibana M (1997) Ca^{2+} activated K^+ current at presynaptic terminals of goldfish retinal bipolar cells. Neurosci Res 27:219–228
13. von Gersdorff H, Matthews G (1996) Calcium-dependent inactivation of calcium current in synaptic terminals of retinal bipolar neurons. J Neurosci 16:115–122
14. Mennerick S, Zenisek D, Matthews G (1997) Static and dynamic membrane properties of large-terminal bipolar cells from goldfish retina: experimental test of a compartment model. J Neurophysiol 78:51–62

Asymmetric Temporal Properties in the Receptive Field of Transient Amacrine Cells in Carp Retina

Masahiro Yamada[1], Masanori Iwasaki[1], Tetsuo Furukawa[2], Syozo Yasui[2], and Kaj Djupsund[3]

Key words. Conduction velocity, Transient amacrine cells, Receptive field

The speed of signal conduction is essential for the temporal properties of neurons and neuronal networks. We observed highly different conduction velocity within the receptive field of fast ON-OFF type transient amacrine cells that tightly coupled each other, forming a retinal network. Photoresponses of the transient amacrine cells in isolated carp retinae were intracellularly recorded using slit photostimuli (110 or 220 μm wide, 7 mm long), which simplified the estimation of the current flow in their network into a one-dimensional problem. The white light slit was scanned along four axes with 45-degree intervals. The averaged conduction velocity (defined by the distance of slit stimuli divided by time to response peak) was about 50 mm/s. This velocity was directionally asymmetric around the observed receptive field centers, with an average maximum ratio of 1:5 in the recorded velocity axis. The fastest speeds were found in the dorsal area of the receptive fields. The asymmetry was similar in the ON- and OFF-part of the responses, and thus independent of the pathway. Despite this, the spatial decay of the graded-voltage photoresponse amplitude in a field was found to be symmetric, separating the temporal and spatial properties of these responses. Thus, it was suggested that the conduction speed can be an independent factor in processing of directional information within the inner retina.

[1] Department of Production, Information and Systems Engineering, Tokyo Metropolitan Institute of Technology, 6-6 Asahigaoka, Hino, Tokyo 191-0065, Japan
[2] Kyushu Institute of Technology, 680-4 Kawazu, Iizuka, Fukuoka 820-8502, Japan
[3] Department of Neuroscience and Neurology, University of Kuopio, Kuopio, Finland

117

Spatial Asymmetry and Temporal Delay of Inhibitory Amacrine Cells Produce Directional Selectivity in Retina

AMANE KOIZUMI[1], MISAKO TAKAYASU[2], HIDEKI TAKAYASU[3], YUTAKA SHIRAISHI[1], and AKIMICHI KANEKO[1]

Key words. Directional selectivity, Simulation, Amacrine cell

Directional selectivity is a unique function relating to agility that some portion of ganglion cells in the retina fire only for moving light signals with specific direction and speed [1]. Taylor et al. [2] showed that inhibitory synaptic outputs from inhibitory amacrine cells to directional selective ganglion cells are playing a critical role in directional selectivity in the rabbit retina. We previously reported that dendrites of inhibitory amacrine cells have active regenerative properties to propagate action potentials [3]. γ-Aminobutyric acid releases from the dendrites were driven by action potential propagation into the dendrites. The speed of action potential propagation in dendrites of amacrine cells was approximately 10 m/s, which is one tenth slower than that of an axon. Thus, initiation and propagation of action potentials on amacrine cell dendrites cause a temporal delay in synaptic outputs to ganglion cells. In addition, asymmetric expansion of dendrites of amacrine cells causes asymmetric synaptic outputs to ganglion cells. Here, we established a novel hypothesis that these features of inhibitory amacrine cells might play an important role in forming directional selectivity. In order to clarify the mechanism of directional selectivity, we numerically analyzed the whole retina activity using a recently developed neural-network simulation powered by NEURON, which models each cell's electrophysiological activity. All known experimental facts reported to date are explained in a consistent manner by our hypothesis on the connection of cells that direction-selective ganglion cells are receiving inhibitory synaptic inputs from amacrine cell dendrites with a random spatial asymmetry and a temporal delay.

[1] Department of Physiology, School of Medicine, Keio University, 35 Shinanomachi, Shinjuku-ku, Tokyo 160-8582, Japan
[2] Department of Complex Systems, Future University-Hakodate, 116-2 Kameda-Nakano-cho, Hakodate, Hokkaido 041-8655, Japan
[3] SONY Computer Science Laboratory, 3-14-3 Higashi-gotanda, Shinagawa-ku, Tokyo 141-0022, Japan

References

1. Barlow HB, Hill RM, Levick WR (1964) Retinal ganglion cells responding selectively to direction speed of image motion in rabbit. J Physiol (Lond) 173:377
2. Taylor WR, He S, Levick WR, et al. (2000) Dendritic computation of direction selectivity by retinal ganglion cells. Science 289:2347–2350
3. Yamada Y, Koizumi A, Iwasaki E, et al. (2002) Propagation of action potentials from the soma to individual dendrite of cultured rat amacrine cells is regulated by local GABA input. J Neurophysiol 87:2858–2866

Fast Calcium Imagings of Amacrine Cell Dendrites in Horizontally Sliced Goldfish Retina

Ryosuke Enoki, Taro Azuma, Kenji Iwamuro, Amane Koizumi, and Akimichi Kaneko

Key words. Dendrite, Calcium imaging, Amacrine cell

Amacrine cells are interneurons that make lateral and vertical connections in the inner plexiform layer of the retina. Amacrine cells have no axon and their dendrites function as both presynaptic and postsynaptic sites. GABAergic amacrine cells constitute 80% of the amacrine cell population in goldfish retina and mediate lateral inhibition between neighboring amacrine cells. Their light-evoked responses consist of regenerative action potentials and excitatory postsynaptic potentials (EPSPs). These depolarizing voltage changes induce Ca^{2+} influx into dendrites, which triggers transmitter release from the storage site in the dendrite. Thus, it is crucial to know how and where Ca^{2+} influx is caused in amacrine cells.

We studied the Ca^{2+} dynamics in amacrine cells by 80×80 high-speed cooled-CCD camera at a frame rate of 0.5–2 kHz. In order to record Ca^{2+} responses from the individual dendrites of intact amacrine cells, goldfish retina was horizontally sliced by a vibratome slicer at the middle level of the inner nuclear layer. In this preparation, horizontal expansion of amacrine cell dendrites is well preserved and was visible tangentially under an IR-DIC microscope. We introduced a Ca^{2+} indicator, Rhod-2, into an amacrine cell from the patch pipette in the whole-cell configuration. Depolarization of the soma triggered action potentials. In response to the action potential, the intracellular Ca^{2+} concentration ($[Ca^{2+}]_i$) was increased first in the soma and such an increase propagated along the dendrites. We also observed spontaneous increases of $[Ca^{2+}]_i$, which is thought to be accompanied by action potentials. We have found that action potentials induced increases of $[Ca^{2+}]_i$ along the entire dendrites. These observations suggest that action potentials trigger transmitter release from the whole dendrites.

Department of Physiology, School of Medicine, Keio University, 35 Shinanomachi, Shinjuku-ku, Tokyo 160-8582, Japan

Distribution and Morphological Features of the Retinal Ganglion Cells in Chicks

Jumpei Naito[1] and Yaoxing Chen[2]

Key words. Vision, Avian retina, Neuron, Morphometry, Bird

Introduction

Birds have a highly specialized visual system that produces a unique visual activity, especially in the visuomotor function. Although the avian visual system has been studied from various aspects, relatively few morphological analyses concerning retinal ganglion cells (RGCs) have yielded precise results [1–5]. The reason is that analysis of avian RGCs is difficult and requires patient effort because of the extraordinarily high density of small RGCs and contamination by many displaced amacrine cells that migrate to the ganglion cell layer (GCL) [6]. In this mini article, we state the distribution and morphological features of chick RGCs.

Total Cell Number and Cell Density in the GCL (Fig. 1)

We estimated the number of RGCs and their density in the GCL using Nissl staining, retrograde cell degeneration by axotomy of the optic nerve, and retrograde cell labeling by injections of horseradish peroxidase (HRP) into the optic nerve in posthatching day 1 (P1) and P8 chicks.

The total number of cells in the GCL was estimated to be 6.1×10^6 (P1) and 4.9×10^6 (P8), and the cell density was 14,300 cells/mm^2 (P1) and 10,400 cells/mm^2 (P8) on average from the computer image analysis of the Nissl preparation. Two high-density areas, the central area (CA) and the dorsal area (DA), were observed in the central and dorsal retinas in both P1 (22,000 cells/mm^2 in CA, 19,000 cells/mm^2 in DA) and P8 (19,000 cells/mm^2 in CA, 12,800 cells/mm^2 in DA) chicks. However, the cell density decreased more rapidly in the DA (67%) than in the CA (86%) by P8. On the other hand, the cell densities in the temporal (TP) and the nasal (NP) periphery

[1] Department of Animal Science, School of Sciences and Engineering, Teikyo University of Science and Technology, Uenohara, Kitatsuru-gun, Yamanashi 409-0193, Japan
[2] Laboratory of Veterinary Anatomy, College of Animal Medicine, China Agricultural University, Beijing 100094, China

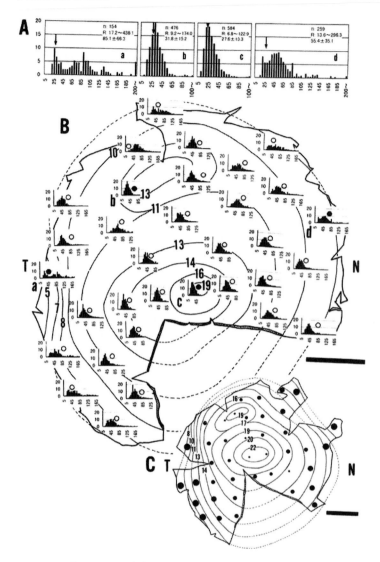

FIG. 1. **A** Frequency histograms of the cell size on the temporal periphery (*a*), dorsal area (*b*), central area (*c*), and nasal periphery (*d*), which are indicated by the same letters (a, b, c, d) in **B**. The *abscissa* and *ordinate* represent the cell size (μm²) and the frequency (%), respectively. *Small arrows* in the histograms indicate the conceivable boundary of the cell sizes between resistant cells and presumptive ganglion cells. *n*, sample number; *R*, range of cell size. Means ± standard deviation of cell sizes. **B** Mosaic of histograms displayed on the isodensity map to show the frequency of cell sizes in the positions indicated by *open or solid circles* [posthatching day 8 (P8)]. The *numbers on the isodensity lines* indicate the cell density (×10³/mm²). N = nasal; T = temporal. *Bar* 5 mm. **C** Averages of cell sizes in each area are represented on the isodensity map (P1) according to the size of circles that are grouped into five ranges: <25 μm², 25–29.9 μm², 30–34.9 μm², 35–39.9 μm², ≥40 μm². *Bar* 5 mm. (From [7], with permission)

were 7800 cells/mm^2 and 12,500 cells/mm^2 in P1 chicks, respectively, and 5000 cells/mm^2 and 8000 cells/mm^2 in P8 chicks, respectively. The cell density in the temporal periphery was about 35% lower than in the nasal periphery in both P1 and P8 chicks [7].

Resistant Cells in the GCL

Thirty percent (1.9 × 10^6 cells in P1) of the total number of cells in the GCL were resistant to axotomy of the optic nerve. These cells must be the displaced amacrine cells and the glia cells. Most of the resistant cells were immunoreactive to the anti-syntaxin antibody. The distribution of the axotomy-resistant cells showed two high-density areas in the CA (5800 cells/mm^2) and the DA (3200 cells/mm^2). These cells also exhibited a gentle center-peripheral gradient (2200 cells/mm^2 in the TP) in P1 chicks.

Cell Sizes and the Cell Number of Presumptive RGCs

The HRP-labeled cells were small in the CA (mean ± SD, 35.7 ± 9.1 μm^2) and DA (40.0 ± 11.3 μm^2), and their sizes increased toward the periphery (63.4 ± 29.7 μm^2 in the TP) accompanied by a decrease in the cell density. However, the axotomy-resistant cells did not significantly increase in size toward the peripheral retina (12.2 ± 2.2 μm^2 in the CA, 15.2 ± 3.2 μm^2 in the DA, 15.1 ± 3.8 μm^2 in the TP). We considered from these cell sizes that presumptive ganglion cells would be larger than 12 μm^2 in the CA, and also larger than 15 μm^2 in the DA and the peripheral retina. Consequently, the total number of presumptive ganglion cells was estimated to be 4 × 10^6 (8600 cells/mm^2 on average), and the density in each area was 13,500 (CA), 10,200 (DA), and 4300 (TP) cells/mm^2 in P8 chicks [7].

The DA has been reported in aerial birds such as pigeons and hawks [1, 8] in addition to the CA, but its existence has not been reported in chicks that live on the ground (terrestrial bird). As described earlier, however, our posthatching study demonstrated the DA that was recognized in P1 chicks, but it was barely detectable in P8 chicks [7].

Types of RGCs in the Central Retina (Fig. 2)

In the CA, we examined the cell type of chick RGCs using retrograde labeling with carbocyanine dye (DiI) and intracellular filling with Lucifer Yellow. Ganglion cells were divided into six groups, Group Ic/Is, IIc/IIs, IIIs, IVc, according to somal size, dendritic branching pattern, and size of the dendritic field. Group I cells (7.3%) had a small somal area and a small dendritic field. They were further divided into two subgroups by complexity (subgroup Ic) and simplicity (subgroup Is) of the dendritic arborization. Group II cells (20.8%) had a medium-sized soma and dendritic field. They were also subdivided into Groups IIc and IIs. Group III cells composed one subgroup (subgroup IIIs) (4.2%). They had medium-sized soma and large and simple dendritic arborization. Group IV cells (2.1%) also composed a single subgroup (subgroup IVc) in which all cells had large soma, and large and complex dendritic arborization [9].

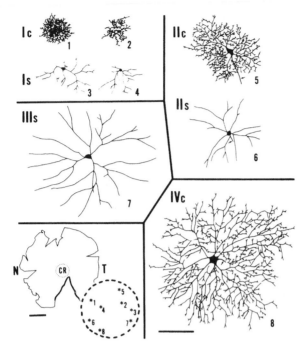

Fig. 2. Camera lucid drawings of retinal ganglion cells (RGCs) classified into six groups (Group Ic/Is, IIc/IIs, IIIs, IVc) in the central retina (CR) shown in the retinal map at the *lower left*. *Arabic numbers* attached to each RGC indicate the position in the central retina represented by the *broken-line circle*. *Bars* represent 5 mm for the retinal map and 100 μm for all RGCs, respectively. (From [9], with permission)

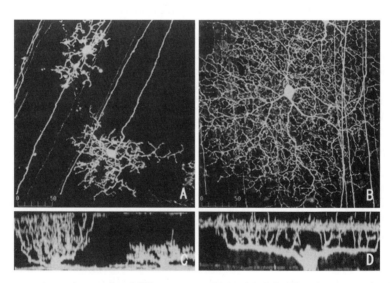

Fig. 3A–D. Photomicrographs of different types of RGCs labeled with carbocyanine dye (DiI), showing the dendritic arborization in the horizontal (**A, B** and vertical **C, D**) planes

Stratification of the Dendritic Arbor in RGCs (Fig. 3)

Other morphological information on RGCs is the dendritic stratification. In a preliminary observation, dendrites of RGCs ramified and formed more than five strata in the inner plexiform layer. All groups contained the RGCs that showed the multidendritic stratification. The relation between the classification of RGCs and stratification of the dendrites was not plain. However, the simple type of RGCs tended to form fewer strata, but the complex type of RGCs displayed various dendritic stratifications.

References

1. Binggeli RL, Paule WJ (1969) The pigeon retina: quantitative aspects of the optic nerve and ganglion cell layer. J Comp Neurol 137:1–18
2. Ehrlich D (1981) Regional specialization of the chick retina as revealed by the size and density of neurons in the ganglion cell layer. J Comp Neurol 195:643–657
3. Ikushima M, Watanabe M, Ito H (1986) Distribution and morphology of retinal ganglion cells in the Japanese quail. Brain Res 376:320–334
4. Wathey JC, Pettigrew JD (1989) Quantitative analysis of the retinal ganglion cell layer and optic nerve of the barn owl *Tyto alba*. Brain Behav Evol 33:279–292
5. Thanos S, Vaneslow J, Mey J (1992) Ganglion cells in the juvenile chick retina and their ability to regenerate axons in vitro. Exp Eye Res 54:377–391
6. Layer PG, Vollmer G (1982) Lucifer yellow stains all displaced amacrine cells of the chicken retina during embryonic development. Neurosci Lett 31:99–104
7. Chen Y, Naito J (1999) A quantitative analysis of cells in the ganglion cell layer of the chick retina. Brain Behav Evol 53:75–86
8. Fite KV, Rosenfield-Wessels S (1975) A comparative study of deep avian foveas. Brain Behave Evol 12:97–115
9. Chen Y, Naito J (1999) Morphological classification of ganglion cells in the central retina of chicks. J Vet Med Sci 61:537–542

Homotypic Gap Junction Connections Between Retinal Amacrine Cells

Soh Hidaka and Ei-ichi Miyachi

Summary. Network properties of gap junction connections between retinal amacrine cells in the inner plexiform layer (IPL) were evaluated by techniques of intracellular recording, dye injection, and electron microscopy conducted on isolated retinas of cyprinid fish. First, amacrine cells were identified with their light-evoked responses to light flashes, either transient ON-OFF or sustained. Second, tracer-coupled networks of the cell populations were visualized by transfer of intracellularly injected Neurobiotin into neighbors. Contacts between the cells were then investigated with high-voltage as well as with conventional electron microscopy. Cell-type specific, homotypic connections in the populations were found. Dendrodendritic contacts were seen either in a tip- or a cross-contact manner. In some experiments, cells belonging to the same cell category (e.g., generating a similar photoresponse profile and expressing similar cell morphology) were stained with other tiny cells, in addition to cells of the same morphology. However, no direct contact between the major and tiny cells was light microscopically identified in these preparations. Electron microscopical analysis revealed gap junctions between light microscopically documented dendrites, but not in dendrites apart from interconnected sites. These results demonstrate that gap junctions between amacrine cells have a homotypical manner in cyprinid fish retinas. The homotypic lateral gap junctions between these cells may play important roles in the inhibitory synaptic behavior in cells in the IPL.

Key words. Gap junction, Homotypic connection, Electrotonic synapse, Inhibitory chemical synapse, Retinal amacrine cell

Introduction

The evidence of gap junction connections [5] between amacrine cells has been accumulating to date (for review, see [14, 23]). Electrophysiologically identified amacrine cells with their characteristic light-evoked responses showed large receptive fields beyond their dendritic arbors [7, 8, 10, 15–19]. Biotinylated molecule injection demon-

Department of Physiology, Fujita Health University School of Medicine, Toyoake, Aichi 470-1192, Japan

strated stereotyped patterns of tracer-coupled networks among bipolar, amacrine, and ganglion cells in a homotypical and/or a heterotypical manner [3, 7, 8, 17, 20–24]. Electron microscopic analysis revealed the presence of anatomically identifiable gap junctions between dendritic or axonal processes of these retinal cells [7, 14, 15, 20]. Thus, an important role of amacrine cells is thought to be the intensification of γ-aminobutyric acid (GABA)ergic synapses among the cell populations in the proximal retina, since lateral gap junction connections between these cells are possible pathways for spreading light-evoked signals laterally in the inner plexiform layer (IPL) via electrotonic transmission [7], and gap junctionally connected amacrine cells contain GABA [12].

Hidaka and Hashimoto [8] summarized groupings of electrophysiologically identified amacrine cells of the Japanese dace by correlation of light-evoked responses to spot illumination with their cellular morphology. Dace amacrine cells were grouped into eight classes. Out of eight classes of dace cells, six different types of tracer-coupling patterns to neighbors were identified. Dendrodendritic contacts in coupled cells were seen either between dendritic tips (tip-contact) or shanks (cross-contact). The response–morphology relationship is compatible to that of carp retina [18, 19]. For investigating lateral pathways for electrotonic transmission between amacrine cells, it is significant to reveal gap junctionally connected networks among these cells. These networks are possible to form syncytium in the IPL and to perform coordinated chemical output synapses onto postsynaptic proximal neurons. The present study was undertaken with two main aims: first, to investigate receptive-field properties of amacrine cells with changes in waveforms of light-evoked responses to various sizes of illumination, and second, to identify gap junctionally connected networks composed by specific cell types. Cells were intracellularly labeled with Neurobiotin following examination of receptive field properties by light stimuli. Histochemical processing revealed dendrodendritic connections between labeled cells. Gap junctions between labeled dendrites were examined by electron microscopy. The present study revealed the occurrence of several types of homotypic gap junction networks in the IPL, but not in a heterotypical manner.

Materials and Methods

The results described as follows were obtained from amacrine cells in the retinas of Japanese dace (*Tribolodon hakonensis*), carp (*Cyprinus carpio*), and goldfish (*Carassius auratus*). Retinas were isolated under anesthesia with MS 222 (Sigma, St. Louis, MO, USA) from the fish, and dark adapted for 1–2 h. Intracellular recordings were made from these cells, using methods described elsewhere in detail [7, 8, 12]. The recording micropipettes contained 6% Lucifer Yellow (LY) (Aldrich, Milwaukee, WI, USA) and 6% Neurobiotin (Vector Laboratories, Burlingame, CA, USA) in 0.5 M LiCl and 0.05 M Tris buffer, pH 7.5. For measurement of the cells' receptive field properties, stimuli of concentric spots and annuli with degrees of light intensity (0.145 mW·cm^{-2} at 0.0 log unit) was delivered. After the receptive field of an individual cell had been examined, the cell was sequentially injected with LY and Neurobiotin by passing polarized currents (±1 nA, duration of 500 ms at 1 Hz for 2 to 10 min). Retinas were fixed in a mixture of 4% paraformaldehyde and 0.1 M phosphate buffer

(PB), pH 7.4. Following observation of LY labeling of the recorded cells, localization of Neurobiotin was visualized by incubation with a solution of avidin-biotin-HRP complex (ABC, Elite kit; Vector Laboratories) and with diaminobenzidine (DAB; Dozin, Tokyo). The DAB visualization was performed by the heavy metal intensification of an HRP reaction in the presence of 0.1 M $(NH_4)_2Ni(SO_4)_2$ and 0.1 M $CoCl_2$, dissolved in 0.1 M Tris buffer, pH 7.8 [1]. The reaction products were further processed by photochromic intensification with 0.02% nitro blue tetrazolium (Sigma) and dissolved in 0.1 M Tris buffer, pH 8.2, for 1 to 10 min under green excitation fluorescent microscopy [22].

For ultrastructural examination of the connections between amacrine cells, we postfixed the tissues in 1% OsO_4 in PB, dehydrated them in a series of ethanol-water mixtures, and embedded them in a mixture of Epoxy resins (Glycidether 100, Serva, Heidelberg, Germany). For enhancement of electron density of the specimens, en bloc stainings were made in three steps of 3% $K_2Cr_2O_7$ in D.W., 2% uranyl acetate in 70% ethanol, and 20% phosphotungstic acid in absolute ethanol. Serial tangential sections (5-μm thickness) of the cells to be studied were made, and the material was examined in a HITACHI H-1250M high-voltage electron microscope at 1,000 kV (Research #02-405: National Institute for Physiological Sciences, Okazaki, Japan). For analysis of junctional structures between the interconnected dendrites, serial ultrathin sections from the specimens were studied in a HITACHI H-7000 or JEOL 1200EX conventional electron microscope at 75 kV or 80 kV fitted with a goniometer stage.

Results and Discussion

Among the classes of fish amacrine cells [8], six different morphological classes of cells were examined. Results were obtained from cells in retinas of Japanese dace ($n = 119$), carp ($n = 33$), and goldfish ($n = 21$). In the present study cells were encountered that were generating either of two main categories of response manners to flashes of light: transient ON-OFF and sustained, when dark-adapted retinas were illuminated under weak intensity at −3.0 log unit. Among the three types of tip-contact cells [8, 12], however, we found that cells expanding their monostratified dendrites in either sublamina (distal or proximal part) of the IPL also generated transient ON-OFF responses to stimuli under a relatively high intensity of illumination (more than −2.0 log unit), although the cells had been classified as the sustained type, based on their light-evoked responses under weak intensity [7, 8, 12, 18, 19].

Figure 1 shows changes in waveform of transient- (Fig. 1A) and sustained-type responses (Fig. 1B), obtained from two different tip-contact cells, respectively. In Fig. 2A,B and 2C,D, individual cells shown in Fig. 1A,B, and their hexagonally tracer-coupled neighbors are indicated, respectively. The cell in Fig. 2A generated typical transient ON-OFF responses to every stimulus (Fig. 1A). In the present study, the transient ON-OFF, tip-contact cells expanding their dendrites in both sublaminas (the distal and proximal parts) of the IPL were most frequently encountered ($n = 47$ in dace). This type of cell was first described in goldfish retina by Kaneko [9, 10] and Kaneko and Hashimoto [11]. Later the transient cells in carp retina, defined as "fast ON-OFF" and "Fnd" type, showed LY coupling in the cell population [18, 19]. The cells may correspond to "A26," "A27," or "A28 type" in roach retina [4]. For the cells in dace,

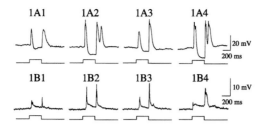

FIG. 1A,B. Intracellular recordings of flash-evoked responses from two types of amacrine cells. A Transient-type responses from a tip-contact bistratified cell, evoked by spots 1 mm (*1A1*) and 2 mm (*1A2*) in diameter, an annulus with a 3.0-mm inner diameter and a 5.0-mm outer diameter (*1A3*), and a field stimulus 8 mm in diameter (*1A4*). Light intensity was maintained at −3.0 log unit. The cell's response to every stimulus was a transient depolarization at the ON or OFF set. Note the increase of a DC hyperpolarization during steady illumination with increasing stimulus size. B Sustained-type response giving a transient depolarization at the cessation of the light from a tip contact distally monostratified cell, evoked by a spot 2 mm in diameter under −3.0 log unit (*1B1*). Transient depolarizations at the ON set were not seen. The cell, however, produced large transient depolarizations at the ON or OFF set to the spot (*1B2*) and annulus under −1.0 log unit (*1B3*). The ON-transient component was not produced in response to the field illumination at −1.0 log unit (*1B4*). Note the absence of a DC hyperpolarization during steady illumination

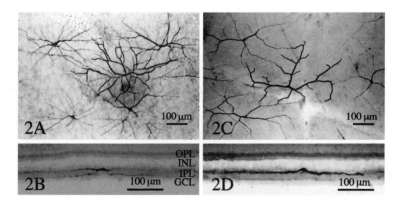

FIG. 2A–D. Cellular morphology and tracer coupling of amacrine cells shown in Fig. 1A,B. Transient cells shown in Fig. 1A are hexagonally connected at dendritic tips with neighbors of the same type (**A**), extending their dendrites at both sublaminas (the distal and proximal layers) of the inner plexiform layer (IPL) (**B**). Cells shown in Fig. 1B are also hexagonally tracer coupled at the tip-contact with neighbors of the same type (**C**), extending their dendrites at the distal part of the IPL (**D**). Note no labeling of cells of other morphological types, apart from the tip-contact homotypic cells in **A** and **C**

cell density, nearest neighbor distance, mean dendritic arbors, and coverage factor were 17.66 cells/mm^2, 148.80 ± 27.11 μm, 0.10 mm^2, and 1.77, respectively. The present study demonstrated the homotypic connection at the dendritic tips among the cell population, since no cellular labeling in other morphological types of cells was seen (Fig. 2A).

The cell in Fig. 2C generated other transient ON-OFF responses to spot and annuli illumination under higher intensity (Fig. 1B2,3), whereas it produced a transient component at the cessation of the spot under relatively weak illumination (Fig. 1B1). The OFF transient component was also seen under the high intensity of the field illumination (Fig. 1B4). The homotypic tip-contact monostratified cells should have been classified into sustained hyperpolarizing of "Fna" [18, 19], based on the spot response [8]. The morphology corresponds to "A6" or "A7 type" of roach [4]. We were not easily able to distinguish these cells (n = 23) with their "online" electrophysiological properties, from the transient-type cell shown in Fig. 2A,B, although the responses shown in Fig. 1B were different from those of Fig.1A in the two components that follow. First, the ON-depolarization was generated gradually, and the amplitude and duration were smaller than that of the OFF transient. Second, during light illumination, a DC depolarization occurred, whereas a DC hyperpolarization was seen in the cell shown in Fig. 1A. "Offline" analysis under microscopical observation revealed that transient ON-OFF responses were from the tip-contact monostratified cells, but not in the bistratified cells. For the cells in dace, cell density, nearest neighbor distance, mean dendritic arbors, and coverage factor were 21.10 cells/mm^2, 215.08 ± 28.03 μm, 0.065 mm^2, and 1.37, respectively. The tip-contact interstitial cells (n = 24) [7], compatible to "Fnb" [18, 19] and "A6/A7" [4], also produced transient ON-OFF to annulus illumination (see Fig. 2A3 in [8]). We here propose, in the study of amacrine cells, that the morphological definition of the cell type should be done, rather than the electrophysiological classification of response type (i.e., in a transient or sustained manner).

Cells with a cross-contact (shank-to-shank) manner were distributed in at least three different ways, corresponding either to the photoresponse type of transient ON-OFF, sustained depolarizing, or hyperpolarizing (not shown). Typical sustained-type responses [9, 10, 16, 18, 19] are known to be generated from the cross-contact monostratified cells [7, 8]. These cells were also homotypically connected, but not distributed in a regular hexagonal array, since neighbors were distributed within individual dendritic arbors [7, 8]. Analysis of the array did not reveal a total cell population. It is, however, possible that several homotypic populations are present with the dendritic arbor size of differential cells even in one category of the sustained depolarizing cross-contact group [18, 19].

In some preparations, tiny amacrine cells were also labeled, in addition to cells of the major homotypic type, as previously reported in goldfish (see Fig. 26 in [23]), but no direct contact between the major and tiny cells was found (not shown). Apparent heterotypic coupling between these cells possibly results from dye leakage or double injection by microelectrode penetration. The tip-contact monostratified cell shown in Fig. 2C made 25 dendrodendritic contacts with homotypic neighbors. Among 31 dendritic tips from one cell, tip-to-tip contacts occurred at 14 tips and the others were found in a tip-to-shank manner. Electron microscopic analysis was performed in the tip contacts by documenting them with their light microscopic

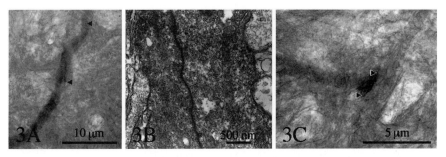

FIG. 3A–C. High-voltage and conventional electron micrographs of gap junctions at Neurobiotin-labeled dendritic tips of amacrine cells. **A** A gap junction (*between arrowheads*) at tip-to-tip contact between dendritic tips of the transient ON-OFF cells shown in Fig. 2A, revealed at 1,000 kV in a tangential 5-μm section in thickness. Note that gap junctions occur between laterally connected dendritic tips. **B** A gap junction at tip-to-tip contact of cells shown in Fig. 2C, observed at 80 kV in an ultrathin section. **C** A small gap junction (*between arrowheads*) at tip-to-shaft contact of transient cells, revealed by high-voltage electron microscopy

images. Figure 3 shows anatomically identified gap junctions laterally localized between Neurobiotin-labeled dendritic tips. High-voltage electron microscopy revealed that the size of gap junction plaques at tip-to-tip contacts was estimated to be 7 μm in the longitudinal axis of the bistratified cells and 16 μm in the monostratified cells, respectively, in a 5-μm thick section (Fig. 3A). The smaller gap junctions of the bistratified cells appeared to occur between more slender dendritic tips, compared with those of the monostratified cells (Fig. 1C,D). The bistratified cells, however, made dual-tip contacts with neighbors in both sublaminas of the IPL. At tip-to-shank contacts, small gap junctions were found (Fig. 3C). The ultrathin sectioning analysis revealed gap junctions comprising seven-layered components [6] at the tip-contacts (Fig. 3B). For cross-contact cells, large gap junctions were found between light-microscopically documented dendritic shanks (not shown). Apart from labeled dendritic contacts between homotypic cells, no gap junction was found in serial-sectioning analysis.

These results demonstrate that fish amacrine cells are gap junctionally connected with each other in the homotypic cell population, but not for morphologically differential cells. The occurrence of a cell's extensive receptive field beyond the dendritic arbors shows that lateral interaction in the IPL appears to be present in the homotypic gap junction connections between amacrine cells. In rabbit, on the other hand, tracer-coupled networks in amacrine cells [24] were reported to fail to extend a cell's receptive field [2]. In mammals, the length of laterally extensive dendrites of amacrine cells was proposed to be important for the extent of spreading the visual signals laterally in the IPL [13]. The present study of fish retinas suggests that lateral electrotonic interaction between homotypic amacrine cells is possible to enforce GABAergic synapses onto proximal retinal neurons.

Acknowledgments. This study was supported by Grants-in-Aid for Scientific Research from the Japanese Ministry of Education, Science and Culture (No. 03857020,

04857016, and 10680754 to S.H.) and by the Science Research Promotion Fund from the Promotion and Mutual Aid Corporation for Private Schools of Japan (No. 231016).

References

1. Adams JC (1981) Heavy metal intensification of DAB-based HRP reaction product. J Histochem Cytochem 29:775.
2. Bloomfield SA (1992) Relationships between receptive field and dendritic field size of amacrine cells in the rabbit retina. J Neurophysiol 68:711–725
3. Dacey DM, Brace S (1992) A coupled network for parasol but not midget ganglion cells in the primate retina. Vis Neurosci 9:279–290
4. Djamgoz MBA, Downing JEG, Wager H-J (1989) Amacrine cells in the retina of a cyprinid fish: functional characterization and intracellular labelling with horseradish peroxidase. Cell Tissue Res 256:607–622
5. Goodenough DA, Goliger JA, Paul DL (1996) Connexins, connexons, and intercellular communication. Annu Rev Biochem 65:475–502
6. Hidaka S, Shingai R, Dowling JE, et al. (1989) Junctions form between catfish horizontal cells in culture. Brain Res 498:53–63
7. Hidaka S, Maehara M, Umino O, et al. (1993) Lateral gap junction connections between retinal amacrine cells summating sustained responses. Neuroreport 5:29–32
8. Hidaka S, Hashimoto Y (1995) Receptive-field properties of retinal amacrine cells in homotypic gap junction networks. In: Kanno Y, Kataoka K, Shiba Y, et al. (eds) Intercellular communication through gap junctions. Progress in cell research, vol. 4. Elsevier Science, Amsterdam, pp 261–264
9. Kaneko A (1970) Physiological and morphological identification of horizontal, bipolar, and amacrine cells in goldfish retina. J Physiol (Lond) 207:623–633
10. Kaneko A (1973) Receptive field organization of bipolar and amacrine cells in the goldfish retina. J Physiol (Lond) 235:133–153
11. Kaneko A, Hashimoto H (1969) Electrophysiological study of single neurons in the inner nuclear layer of the carp retina. Vision Res 9:37–55
12. Lu Y, Hidaka S, Hashimoto Y (1995) GABAergic inhibitory synapses from amacrine cells in the fish retina: evidence for GABA immunoreactivity in amacrine cells connected with tip-contact dendritic gap junctions. J Tokyo Wom Med Coll 65:108–122
13. Masland RH (2001) The fundamental plan of the retina. Nat Neurosci 4:877–885
14. Miyachi E, Hidaka S, Murakami M (1999) Electrical couplings of retinal neurons. In: Toyoda J, Murakami M, Kaneko A, et al. (eds) The retinal basis of vision. Elsevier Science, Amsterdam, pp 171–184
15. Naka KI, Christensen B (1981) Direct electrical connections between transient amacrine cells in the catfish retina. Science 214:462–464
16. Sakai HM, Naka KI (1992) Response dynamics and receptive-field organization of catfish amacrine cells. J Neurophysiol 67:430–442
17. Sakai HM, Naka K-I (1997) Processing of color- and noncolor-coded signals in the gourami retina: II. Amacrine cells. J Neurophysiol 78:2018–2033
18. Teranishi T, Negishi K, Kato S (1987) Correlations between photoresponse and morphology of amacrine cells in the carp retina. Neuroscience 20:935–950
19. Teranishi T, Negishi K (1994) Double-staining of horizontal and amacrine cells by intracellular injection with lucifer yellow and biocytin in carp retina. Neuroscience 59:217–226.
20. Umino O, Maehara M, Hidaka S, et al. (1994) The network properties of bipolar-bipolar coupling in the retina of teleost fishes. Vis Neurosci 11:533–548

21. Vaney DI (1991) Many diverse types of retinal neurons show tracer coupling when injected with biocytin or Neurobiotin. Neurosci Lett 125:187–190

22. Vaney DI (1992) Photochromic intensification of diaminobenzidine reaction product in the presence of tetrazolium salts: applications for intracellular labelling and immunohistochemistry. J Neurosci Methods 44:217–223

23. Vaney DI (1994) Patterns of neuronal coupling in the retina. In: Osborne NN, Chader GJ (eds) Progress in retinal eye research, vol. 13. Pergamon Press, Oxford, pp 301–355

24. Xin D, Bloomfield SA (1997) Tracer coupling pattern of amacrine and ganglion cells in the rabbit retina. J Comp Neurol 383:512–528

A Retinal Ganglion Cell Model Based on Discrete Stochastic Ion Channels

Yoshimi Kamiyama[1] and Shiro Usui[2]

Key words. Retinal ganglion cell, Ion channel, Markov process, Spike generation, Mathematical model

Introduction

The ganglion cells of the vertebrate retina form the pathway by which the retina communicates with the visual cortex. The ganglion cells convert the graded potentials into a pattern of spikes whose characteristics are modulated by the synaptic and membrane currents. Voltage-clamp studies of retinal ganglion cells have identified voltage- or ion-gated currents, which appear to play a role in generating spikes [1]. In previous studies, the ionic conductances have been modeled by means of deterministic differential equations similar to the Hodgkin-Huxley formulation [2, 3]. Recently, however, it was suggested that the stochastic properties of ionic channels are critical in determining the reliability and accuracy of neuron firing [4]. It is important, therefore, to clarify the relationship between membrane excitability and channel stochastics in retinal ganglion cells.

We propose here a stochastic model of spike generation in retinal ganglion cells, based on discrete stochastic ion channels represented by Markov processes. Voltage-clamp simulations show that, as the size of membrane area increases, the response from the stochastic model shows a similar behavior predicted by the deterministic equations. Current clamp simulations show that the reliability and accuracy of spike patterns are highly correlated with the characteristics of the input current. The results suggest that the stochastic properties of ion channels play an important role in determining the firing patterns of retinal ganglion cells.

[1] Information Science and Technology, Aichi Prefectural University, 1522-3 Ibaragabasama, Kumabari, Nagakute, Aichi 480-1198, Japan
[2] Information and Computer Sciences, Toyohashi University of Technology, 1-1 Hibarigaoka Tempaku, Toyohashi, Aichi 441-8580, Japan

Methods

The macroscopic membrane dynamics are modeled by five ionic currents: voltage-gated sodium current (I_{Na}), calcium current (I_{Ca}), transient outward current (I_A), Ca-activated current ($I_{K(Ca)}$), and the delayed rectifying potassium current (I_K) [3]. The membrane potential V is given by

$$C\frac{dV}{dt} = I_{ext} - (I_{Na} + I_{Ca} + I_A + I_{K(Ca)} + I_K + I_L)$$
$$= I_{ext} - \{g_{Na}(V - V_{Na}) + g_{Ca}(V - V_{Ca}) + g_A(V - V_K)$$
$$+ g_{K(Ca)}(V - V_{K(Ca)}) + g_K(V - V_K) + g_L(V - V_L)\}$$

where C is membrane capacitance; I_{ext} is externally applied current; and I_L is leakage current.

In this study, we developed a stochastic model of retinal ganglion cells, based on discrete stochastic ion channels represented by n-state Markov processes [5], i.e., Na channel by eight states, Ca channel by four states, A channel by eight states, and K channel by 5 states. We assumed that K(Ca) and leakage channels do not exhibit stochastic characteristics in the present model. Figure 1 shows the Markov kinetic scheme of a Na channel. Each channel is assumed to randomly fluctuate between stable states ($[m_i h_j]$). Transition probabilities between these states are assumed to depend on the present state and the present membrane potential. The voltage dependent rate constants from the deterministic model [3] were used to estimate the transition probabilities. To simulate the channel behavior, we set random numbers for the initial state that the channel occupies, the duration of the state, and the subsequent state in each time step.

Results

Figure 2 shows the voltage-clamp step response of the membrane conductances, g_{Na}, g_K, g_A, and g_{Ca}. The membrane potential is held at $-65\,$mV and then depolarized in a step to $20\,$mV. The top traces in Fig. 2 represent the responses from the deterministic equations. The responses of five different size membrane patches were simulated. As the membrane size is increased, the continuous deterministic behavior of each conductance emerges from the stochastic single-channel behavior.

Figure 3 shows the simulated firing patterns to DC and Gaussian current injections in the deterministic and stochastic models. The deterministic model shows a train of regular firing (Fig. 3, middle panel). The response of the stochastic model varies from trial to trial, and thus ten superimposed responses to the same input current are

FIG. 1. Markov kinetic scheme of a Na channel, $[m_3 h_1]$ is the open state of the channel and conducts ions. α_m, β_m, α_h, and β_h are the voltage-dependent rate functions from the deterministic model [3]

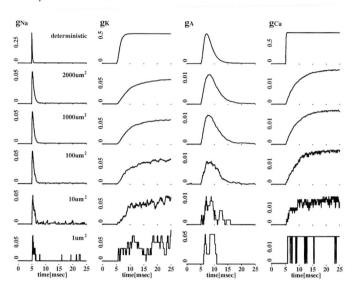

FIG. 2. Voltage-clamp response of the membrane conductances, g_{Na}, g_K, g_A, and g_{Ca}. The membrane potential is held at −65 mV and then depolarized to 20 mV. The *top trace* represents the deterministic responses. The *second and the rest of the traces* are the responses by the stochastic model with varying the membrane size of 1 to 2000 μm². Na, K, A, and Ca channel densities are assumed to be 25, 6, 18, and 1.1 μm⁻², respectively. All single channel conductances are assumed to be 0.02 nS

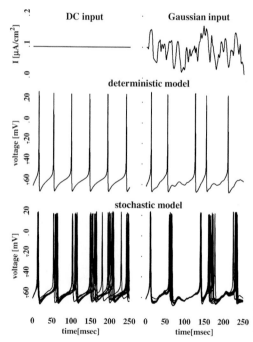

FIG. 3. Firing patterns of the deterministic and stochastic models in response to DC and Gaussian fluctuating current injections. Ten responses to the same input are superimposed for the stochastic model. Membrane patch area used was 490 μm²

shown (Fig. 3, lower panel). In contrast to the case with DC pulse responses, when the stimulus is randomly fluctuating, the stochastic model shows relatively precise and stable spike timing.

Discussion

The stochastic model of the retinal ganglion cell based on discrete ionic channels has been compared with the deterministic model based on the conventional Hodgkin-Huxley-type equations. Under current clamp conditions, the firing patterns simulated by the stochastic model showed a dramatic difference from those of the deterministic model. The predicted spiking characteristics should be demonstrated experimentally.

In the present study, the ganglion cell was modeled by five ionic currents. However, more than a dozen voltage-, ion-, and ligand-gated ionic channels have been identified in retinal ganglion cells [1]. A more realistic model of the spike-encoding mechanisms, which takes into account various ionic channels, as well as dendritic morphologies, should be developed.

Acknowledgments. This work was supported in part by the Grant-in-Aid for Scientific Research (C) (No. 14550424) from the Japan Society for the Promotion of Science, and the Project on Neuroinformatics Research in Vision through Special Coordination Funds for Promoting Science and Technology from the Ministry of Education, Culture, Sports, Science and Technology, the Japanese Government.

References

1. Ishida AT (1995) Ion channel components of retinal ganglion cells. Prog Retin Eye Res 15:261–280
2. Hodgkin AL, Huxley AF (1952) A quantitative description of membrane current and its application to conduction and excitation in nerve. J Physiol 117:500–544
3. Fohlmeister JF, Miller RF (1997) Impulse encoding mechanisms of ganglion cells in the tiger salamander retina. J Neurophysiol 78:1935–1947
4. Schneidman E, Freedman B, Segev I (1998) Ion channel stochasticity may be critical in determining the reliability and precision of spike timing. Neural Computat 10:1679–1703
5. Clay JR, DeFelice LJ (1983) Relationship between membrane excitability and single channel open-close kinetics. Biophys J 42:151–157

Retinal Adaptation to the Spatiotemporal Statistical Structure of Visual Input

Toshihiko Hosoya and Markus Meister

Key words. Ganglion cells, Adaptation, Statistical structure

Under natural conditions, the statistical characteristics of sensory input vary considerably. For efficient representation, the sensory systems should adjust their coding rules. The best understood example is retinal adaptation to the mean light level. Recently it was also found that retina adapts to the contrast, the range of the light level variation. Visual input is not a single variable, but composed of many "pixels", or multiple variables. Even when the mean and the contrast of these variables are constant, the statistical relationship among the variables can vary. Theory predicts that it is again advantageous to adjust coding rules. Since previous experiments suggested that the retinal behavior changes according to such statistical structures, we investigated the class of statistical structures that lead to response adjustments, and the profiles of those adjustments.

The receptive fields of salamander ganglion cells were divided into several regions. For adaptation, the brightness of these regions was modulated according to various statistical structures, keeping the mean and the contrast constant. After adaptation, the response linear kernels were measured using white noise stimuli to characterize the spatiotemporal sensitivity of the cells.

Interestingly, the ganglion cell responses showed significant alterations. The alterations were seen after adaptation to a variety of spatiotemporal statistical structures, including purely spatial correlations and complex spatiotemporal correlations. Different adaptation stimuli caused different response modifications. In general, the cells become more sensitive to deviation from the statistical structure used for adaptation.

Thus, the retina measures the statistical relationship within the input and adjusts the coding rule. The adaptation profiles suggest that the adaptation can function for efficient coding and novelty detection. Since this adaptation occurs for a broad class of statistical structures, it might be a basic design principle of the retina. Possible underlying neural mechanisms involving anti-Hebbian synapses will be discussed.

Department of Molecular and Cellular Biology, Harvard University, Cambridge, MA 02138, USA

Zinc as a Neurotransmitter in the Rat Retina

M. Kaneda[1], K. Ishii[2], T. Akagi[2], and T. Hashikawa[2]

Key words. Zinc, Retina, Electron microscopy, Patch clamp, GABA

Introduction

In the retina it is hypothesized that zinc is a neurotransmitter. Here, we describe the histological localization of zinc, the biological actions of zinc, the uptake of zinc, and the release of zinc based on our recent findings in the rat retina.

Localization of Zinc in the Rat Retina

We have studied the subcellular localization of zinc using the silver amplification method [1, 2] in the rat retina [3]. Under the light microscope, reaction products (silver grains) were detected in the pigment epithelial cells, the inner segment of photoreceptors, the outer nuclear layer, the inner nuclear layer, the outer plexiform layer (OPL), the inner plexiform layer (IPL), and the ganglion cell layer (GC) (Fig. 1A). To rule out the contamination of silver grains reacted with other divalent cations, we used diethyldithiocarbamate (DEDTC) pretreatment to chelate zinc [1, 2]. We occasionally observed silver grains associating with the nucleus of photoreceptors after pretreatment, but at a much lower level than in untreated retina (Fig. 1B), suggesting that the distribution of silver grains reflects the presence of zinc in the rat retina.

Under the electron microscope, zinc was associated with neural processes in OPL and IPL (Fig. 2). We could not find any silver grains in the terminals with synaptic ribbons. Our findings indicate that zinc exists in the non-glutamatergic terminals. We are now trying to identify the zinc-containing terminals by double labeling with immunohistochemical markers and silver amplification.

[1] Department of Physiology, School of Medicine, Keio University, 35 Shinanomachi, Shinjuku-ku, Tokyo 160-8582, Japan
[2] Laboratory for Neural Architecture, BSI, RIKEN, 2-1 Hirosawa, Wako, Saitama 351-0198, Japan

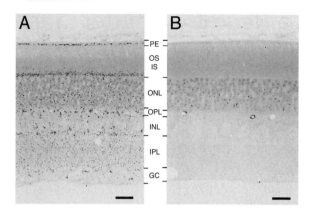

FIG. 1A,B. Photomicrographs showing the distribution of zinc in the rat retina. Zinc was detected as silver grains developed by the IntenSE M silver enhancement kit. **A** Semithin section without diethyldithiocarbamate (DEDTC) pretreatment. **B** Semithin section with DEDTC pretreatment. Thickness of section, 2 μm. *Bars* 20 μm. *PE*, pigment epithelial cells; *OS*, outer segment of photoreceptors; *IS*, inner segment of photoreceptors; *ONL*, outer nuclear layer; *OPL*, outer plexiform layer; *INL*, inner nuclear layer; *IPL*, inner plexiform layer; *GC*, ganglion cell layer. (Reproduced, with permission from Akagi T et al.: Differential subcellular localization of zinc in the rat retina. J Histochem Cytochem 47:87–96, 2001)

Studies of the Biological Actions of Zinc in the Retina

In the rat retina, inhibitory actions of zinc on γ-aminobutyric acid (GABA)$_C$ responses are complete at >10 μM, while inhibition of GABA$_A$ receptors by zinc did not occur at 10 μM in bipolar cells of the mouse retina [4] (Fig. 3). It has been estimated that the concentration of Zn^{2+} at the synaptic cleft would be ~3 μM, when synaptic vesicles containing Zn^{2+} are released into the synaptic cleft [5]. This calculation may mean that GABA$_C$ receptors are the physiological target of Zn^{2+}. Our recent histochemical findings in the rat retina show that non-glutamatergic terminals contain zinc [3]. We are now studying the biological actions of zinc on glycine responses.

Release of Zinc from Retina

We measured by atomic absorption spectrophotometry whether zinc is released from the rat retina. We occasionally observed an increase in zinc concentration of about 1 ppb by high K stimulation. This increase was smaller than the contaminated zinc (2–3 ppb), suggesting difficulties for precise measurement of released zinc.

Uptake of Zinc in the Retina

Currently there is no report in the retina. The zinc uptake mechanism should be studied in future experiments.

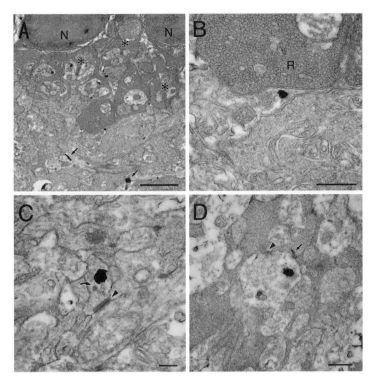

Fig. 2A–D. Zinc found in the synaptic terminals of OPL (**A, B**) and IPL (**C, D**). **A** Rod spherules were characterized by the synaptic ribbons (*asterisks*). Silver grains are also seen in the light neural processes without any contact with rod spherules (*arrows*). *N*, nuclei of the photoreceptors. **B** The light neural process contacting rod spherules (*R*). **C** Silver grain is associated with the electron-lucent neural process with a nonsynaptic membrane junction (*arrowhead*) and cored vesicles (*curved arrow*). **D** Silver grain is associated with the electron-lucent neural process with many small vesicles. A synaptic density between the electron-lucent neural process and the terminal of the bipolar cell is indicated by an *arrowhead*. Synaptic ribbon is seen in the terminal of the bipolar cell (*arrow*). *Bars* 2 μm for **A**; 0.5 μm for **B–D**. (Reproduced, with permission from Akagi T et al.: Differential subcellular localization of zinc in the rat retina. J Histochem Cytochem 47:87–96, 2001)

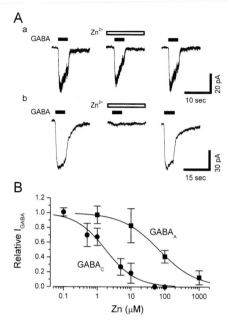

Fig. 3. **A** Inhibitory effects of Zn^{2+} on I_{GABA-A} (Aa) and I_{GABA-C} (Ab). Currents were evoked by 10 μM γ-aminobutyric acid (*GABA*) applied every 3 min in the presence and absence of 10 μM Zn^{2+}. The timing of GABA applications is indicated by *filled bars* and that of 10 μM Zn^{2+} by *open bars*. **B** Concentration-inhibition curve of $[Zn^{2+}]_o$ on I_{GABA-A} and I_{GABA-C}. Each response was normalized to a control response evoked by application of 10 μM GABA in the absence of Zn^{2+}. Each point represents the mean of 3–9 cells and *error bars* indicate SD. Data points were fitted to a Hill equation ($GABA_A$, Hill coefficient of 0.80, IC50 of 67.4; $GABA_C$, Hill coefficient of 0.97, IC50 of 1.8). Vh = −47 mV. (Reproduced, with permission, Modulation by Zn^{2+} of GABA responses in bipolar cells of the mouse retina, Kaneda M et al.: Vis Neurosci 17:273–281, 2000)

Acknowledgments. Electrophysiological studies were carried out with the help of Professor A. Kaneko and Dr. B. Andrásfalvy. Atomic absorption spectrophotometry was carried out with the help of Professor K. Omae and Ms. K. Hosoda.

References

1. Danscher G (1981) Histochemical demonstration of heavy metals. A revised version of the sulphide silver method suitable for both light and electronmicroscopy. Histochemistry 71:1–16
2. Danscher G (1996) The autometallographic zinc-sulphide method. A new approach involving in vivo creation of nanometer-sized zinc sulphide crystal lattices in zinc-enriched synaptic and secretory vesicles. Histochem J 28:361–373
3. Akagi T, Kaneda M, Ishii K, et al. (2001) Differential subcellular localization of zinc in the rat retina. J Histochem Cytochem 49:87–96
4. Kaneda M, Andrásfalvy B, Kaneko A (2000) Modulation by Zn^{2+} of GABA responses in bipolar cells of the mouse retina. Vis Neurosci 17:273–281
5. Frederickson CJ, Suh SW, Silva D, et al. (2000) Importance of zinc in the central nervous system: the zinc-containing neuron. J Nutr 130:1471S–1483S

Suppression by Odorants of Voltage-Gated and Ligand-Gated Channels in Retinal Neurons

Mahito Ohkuma, Fusao Kawai, and Ei-ichi Miyachi

Key words. Odorant, Voltage-gated channel, Cyclic nucleotide-gated channel, Glutamate-gated channel

Introduction

The initial step in olfactory sensation involves the binding of odorant molecules to specific receptor proteins on the ciliary surface of olfactory receptor cells (ORCs). Odorant receptors coupled to G-proteins activate adenylyl cyclase, leading to the generation of cAMP, which directly gates a cyclic nucleotide-gated (CNG) cationic channel in the ciliary membrane [1–5]. In contrast, odorants such as amyl acetate, limonene, and acetophenone (banana-, lemon-, and orange-blossom-like-odor) are also known to suppress the CNG current in ORCs [6]. Similar suppression was observed in various voltage-gated currents of olfactory cells [7–9]. It is thought that suppression influences the perception of odorant by masking odorant responses and by sharpening the odorant specificities in single cells. But it is unclear whether odorants suppress ion channels in ORCs specifically or whether they also suppress channels in other neurons generally. In the present study, we examined the effect of odorants on ionic channels in retinal cells, because they were well-characterized and could be treated easily. Although odorants are not stimulants or endogenous molecules in the retina, it would be interesting to elucidate biophysical and pharmacological properties of odorants against the ionic channels.

Odorants Suppress Voltage-Gated Channels in Retinal Horizontal Cells

It is unclear whether odorants suppress voltage-gated currents in ORCs specifically or whether they also suppress currents in other neurons generally. To investigate this, we examined horizontal cells from the goldfish retina [10]. The major currents measured in goldfish horizontal cells include an L-type Ca^{2+} current (I_{Ca}), a delayed rectifier K^+

Department of Physiology, Fujita Health University, Toyoake, Aichi 470-1192, Japan current (I_K), a fast transient K^+ current (I_A), and an anomalous rectifier K^+ current (I_{Ka}) [11]. Each voltage-gated current was isolated using several pharmacological agents and recorded under whole-cell patch-clamp conditions. As expected, odorants suppressed nonselectively the voltage-gated currents in horizontal cells. Bath application of 10 mM amyl acetate almost completely suppressed the steady inward current and also suppressed the currents evoked by the voltage pulses over the entire voltage range. The current showed full recovery after washout of odorant. Current suppression was also observed by bath application of acetophenone and limonene.

To elucidate the blocking mechanisms by odorants, we measured the time course of the current suppression. Amyl acetate almost completely suppressed I_{Ca} induced by the first step of depolarization during puff application. Thus, binding of an odorant molecule to the open channel is not required to block the I_{Ca}. After cessation of the puff, the I_{Ca} amplitude recovered in 4 s (with a time constant of approximately 600 ms), which was significantly slower than the time resolution (~20 ms) for the puff application system. Similar results were obtained for I_K. The suppressive effects of amyl acetate on I_{Ca} and I_K was immediate, suggesting that binding of an odorant to the open channel is not required to block these channels. Therefore, amyl acetate is a closed channel blocker of I_{Ca} and I_K.

Previous study showed that odorants suppress Ca^{2+} currents in ORCs by making the slope of their inactivation curves gentler and by shifting the half-inactivation voltages toward a negative voltage [9]. To test whether the blocking mechanism of I_{Ca} in horizontal cells is similar, we measured activation and inactivation curves of the horizontal cell's I_{Ca} in the presence and absence of amyl acetate. Although the activation curve was not changed significantly, the inactivation curve significantly shifted its half-inactivation voltage toward a negative voltage by the application of amyl acetate. We also examined the effects of amyl acetate on I_K. Amyl acetate did not change the inactivation curve of I_K significantly. These results indicate that odorants suppress nonselectively the voltage-gated currents in horizontal cells in a similar manner as in ORCs.

Odorants Suppress Cyclic Nucleotide-Gated Channels in Photoreceptor Cells

Cyclic nucleotide-gated current is suppressed by odorants in ORCs. It is unclear, however, whether odorants suppress the CNG current directly or whether they suppress the currents via second messenger [i.e., cyclic adenosine monophosphate (cAMP)] metabolism activated by odorant binding to odorant receptors. To investigate these possibilities, we examined the CNG currents in the newt rod and cone photoreceptor cells, because the photoreceptors do not have the second messenger systems activated by odorants. Thus, by using the photoreceptors, one can examine direct effects of odorants on the CNG channels. In this study, we investigated mechanisms underlying the suppression by odorants of the CNG currents using the whole-cell version of the patch-clamp technique.

We found that odorants also suppress CNG currents in photoreceptor cells. Under voltage clamp, an odorant puff immediately suppressed the currents induced by the intracellular cyclic guanosine monophosphate (cGMP) in isolated newt rods and cones. This result raises two possible mechanisms. First, amyl acetate may reduce the rod CNG current by blocking its channel directly. Second, amyl acetate might reduce the CNG current by decreasing the intracellular cGMP concentration through the activation of a cGMP-hydrolyzing phosphodiesterase. To rule out the latter, we used a nonhydrolyzable cGMP analog, 8-p-chlorophenylthio-cGMP (8-CPT-cGMP). Odorants also suppressed the currents induced by another cGMP analog, suggesting that the second messenger metabolism via phosphodiesterase is not involved in the suppression by odorants. Thus, it is likely that amyl acetate reduces the rod CNG current by blocking its channel directly.

Odorants Suppress Ionotropic Glutamate Receptors in Retinal Neurons

Glutamate-gated channels are expressed in the retinal neurons. To investigate whether odorants can also suppress glutamate-gated channels, we used isolated retinal neurons in amphibians, because the glutamate-gated channels in these cells have been well-characterized. Since Ca^{2+} is permeable through some types of ionotropic glutamate receptors, in this study we examined the effects of odorants on the glutamate-gated channels using the Fura-2-based Ca^{2+} imaging technique [13]. Bath application of glutamate rose $[Ca^{2+}]_i$ under application of the voltage-gated Ca^{2+} channel blocker. Thus, $[Ca^{2+}]_i$ rises in the neurons were most likely attributable to Ca^{2+} influx via Ca^{2+}-permeable glutamate-gated channels rather than voltage-gated Ca^{2+} channels. Immediately following 100 μM glutamate, the neuron was superfused with the solution containing 100 μM glutamate and odorant. As expected, amyl acetate markedly and reversibly reduced $[Ca^{2+}]_i$ in the retinal neuron.

Ionotropic glutamate receptors are classified into N-methyl-D-aspartate (NMDA) receptors and AMPA/kainate receptors. We investigated whether odorants suppress both NMDA receptors and α-amino-3-hydroxy-5-methylisoxazole-4-propionic acid (AMPA) kainate receptors, or suppress selectively each one only. Application of odorants reversibly reduced $[Ca^{2+}]_i$ increased by NMDA and kainate. This suggests that odorants can suppress not only voltage-gated channels but also ligand-gated channels such as NMDA and AMPA/kainate receptors.

Conclusion

We found that odorants also suppress voltage-gated and ligand-gated channels in neurons outside of the olfactory system. The suppressive effects of amyl acetate on I_{Ca} and I_K was immediate in horizontal cells, suggesting that binding of an odorant to the open channel is not required to block these channels. Therefore, amyl acetate is a closed channel blocker of I_{Ca} and I_K. These phenomena are similar to the suppressive effect of local anesthetics (lidocaine and benzocaine) in various preparations. Local anesthetics are known as Na^+ channel blockers. Furthermore, local anesthetics non-selectively suppress not only I_{Na} but also various kinds of voltage-gated currents such as I_K and I_{Ca}. These observations are similar to our results, suggesting that the blocking mechanisms may be similar to those of local anesthetics.

Chemical structures of the odorants used in the present experiment (see Table 1 in [12]) show the high lipid solubility and neutral charge. Thus, it is likely that these molecules can readily cross the cell membranes and can bind to hydrophobic sites of ion channels. In addition, it is known that odorants nonselectively suppress the voltage-gated channels in the olfactory cells [7–9]. Therefore, we suggest that odorants may suppress both voltage-gated and ligand-gated channels in a similar manner as the ion channels in ORCs. However, it is still unclear how the odorant molecules alter the ionic permeation of ion channels in the neurons. It may be worthwhile to investigate the gating kinetics with single-channel recording.

Acknowledgments. This work was supported by Japan Society of the Promotion of Science and the Naito Foundation.

References

1. Bakalyar HA, Reed RR (1991) The second messenger cascade in olfactory receptor neurons. Curr Biol 1:204–208
2. Firestein S, Werblin FS (1987) Gating currents in isolated olfactory receptor neurons of the larval tiger salamander. Proc Natl Acad Sci USA 88:6292–6296
3. Gold GH, Nakamura T (1987) Cyclic nucleotide-gated conductances: a new class of ionic channels mediates visual and olfactory transduction. Trends Pharmacol 8:312–316
4. Kurahashi T, Yau K-W (1994) Tale of an unusual chloride current. Curr Biol 4:256–258
5. Restrepo D, Teeter JH, Schild D (1996) Second messenger signaling in olfactory transduction. J Neurobiol 30:37–48
6. Kurahashi T, Lowe G, Gold GH (1994) Suppression of odorant responses by odorants in olfactory receptor cells. Science 265:118–120
7. Kawai F, Kurahashi T, Kaneko A (1997) Nonselective suppression of voltage-gated currents by odorants in the newt olfactory receptor cells. J Gen Physiol 109:265–272
8. Kawai F (1999a) Odorant suppression of delayed rectifier potassium current in newt olfactory receptor cells. Neurosci Lett 269:45–48
9. Kawai F (1999b) Odorants suppress T- and L-type Ca^{2+} currents in olfactory receptor cells by shifting their inactivation curves to a negative voltage. Neurosci Res 35:253–263
10. Kawai F, Miyachi E (2000a) Odorants suppress voltage-gated currents in retinal horizontal cells in goldfish. Neurosci Lett 281:151–154
11. Tachibana M (1983) Ionic currents of solitary horizontal cells isolated from goldfish retina. J Physiol (Lond) 345:329–351
12. Kawai F, Miyachi E (2000b) Direct suppression by odorants of cyclic nucleotide-gated currents in the newt photoreceptors. Brain Res 876:180–184
13. Ohkuma M, Kawai F, Miyachi E (2002) Direct suppression by odorants of ionotropic glutamate receptors in newt retinal neurons. J Neural Transm 109:1365–1371

Dopamine D_1 Receptor on Cultured Müller Cells

Risako Minei[1], Masato Wakakura[2], and Kimiyo Mashimo[1]

Key words. Müller cells, Dopamine receptor, Intracellular calcium

Aim

Müller cells are multifunctional glial cells, which are recently reported to possess various types of neurotransmitter receptors. It has been reported that dopamine has a neuroprotective effect on the retina. The effect appears to be mediated by dopamine receptor on the retinal cells, but dopamine receptors on Müller cells have not yet been confirmed. To clarify the presence of a nonneurotransmitter effect of dopamine, we studied whether a calcium-mediated metabotropic receptor, dopamine D_1 receptor, is functionally expressed on cultured Müller cells.

Methods

Müller glial cells were cultured from adult rabbit retinas. Intracellular calcium concentrations were measured before and after administration of dopamine D_1 receptor agonist (R(+)-SKF-82957 hydrobromide) under several conditions using cell labeling of a calcium ion indicator, Fura-2AM.

Results

Müller cells responded to D_1 agonist in a dose-dependent manner at concentrations in the range of 10^{-6} and $10^{-3} \mu M$. Under the presence of D_1 antagonist at $10^{-4} \mu M$, the response significantly diminished at the rate of a positive cell number, as well as with an increasing level of intracellular calcium ion. With the depletion of calcium ions in the medium, a positive response could not be detected.

Conclusions

Müller cells cultured from rabbit retina appeared to possess functional dopamine D_1 receptor.

---

[1] Department of Ophthalmology, Kitasato University, School of Medicine, 1-15-1 Kitasato, Sagamihara, Kanagawa 228-8555, Japan
[2] Inouye Eye Hospital, 4-3 Kanda-Surugadai, Chiyoda-ku, Tokyo 101-0062, Japan

Origin of Intracellular Ca^{2+} Concentration Increase by Hypoxia Differs Between Müller Cells and Ganglion Cells

Tsugihisa Sasaki and Akimichi Kaneko

Key words. Ischemia, Müller cell, Ganglion cell, Voltage-gated Ca^{2+} channel, NMDA, Selective loading of Ca-sensitive dyes

It is thought that ischemia-induced neuronal cell death is triggered by calcium influx through the glutamate receptor. However, the intracellular Ca^{2+} concentration ($[Ca^{2+}]_i$) change by ischemic damage has not been studied in individual cell types. We measured $[Ca^{2+}]_i$ change, caused by a hypoxic condition, of Müller cells (MC) and ganglion cells (GC) by selective loading of Ca-sensitive dyes in sliced rat retina.

Injection of 2 μl of 10 mM fura-2 potassium salts into the vitreous of 6-week-old rats selectively stained MC. Application of 40 mM fluo-3 to the cutting edge of the optic nerve selectively stained GC. Ca imaging was captured by ARGUS-50 (Hamamatsu Photonics, Hamamatsu, Japan). A hypoxic condition was applied by 2-deoxy-D-glucose containing an external solution saturated with a 95% N$_2$, 5% CO$_2$ gas mixture.

A 15-min hypoxic condition induced a $[Ca^{2+}]_i$ increase in both MC (Δ ratio: 0.11 ± 0.02, mean ± SD) and GC (0.14 ± 0.04). Application of 60 μM 2-amino-5-phosphonovaleric acid (APV) partially suppressed the $[Ca^{2+}]_i$ increase in both MC (0.08 ± 0.02, $P < 0.05$) and GC (0.09 ± 0.02, $P < 0.005$). Neither 6-cyano-7-nitroquinoxaline-2,3-dione disodium (CNQX) nor DL-TBOA, glutamate transport inhibitors, inhibited the $[Ca^{2+}]_i$ increase in MC and GC. An increase in $[Ca^{2+}]_i$ was not observed in the presence of 10 μM nifedipine (0.02 ± 0.02, $P < 0.005$) or zero Ca^{2+} extracellular solution ($-0.01 ± 0.02$, $P < 0.005$) in MC. In GC, $[Ca^{2+}]_i$ elevation was not inhibited by nifedipine but was partially blocked by a Ca-free solution (0.10 ± 0.04, $P < 0.05$). Bepridil, a Na/Ca exchange blocker, did not suppress the $[Ca^{2+}]_i$ increase in MC and GC.

Our selective staining method revealed that the mechanisms of hypoxia-induced $[Ca^{2+}]_i$ increase differ between MC and GC. In MC, a voltage-gated Ca^{2+} channel mainly participates in the hypoxia-induced $[Ca^{2+}]_i$ increase, while in GC, Ca^{2+} through an NMDA-gated channel is a partial source of the $[Ca^{2+}]_i$ increase.

Department of Physiology, School of Medicine, Keio University, 35 Shinanomachi, Shinjuku-ku, Tokyo 160-8582, Japan

Expression of Otx2 in Regenerating and Developing Newt Retina

Sanae Sakami, Osamu Hisatomi, Shunsuke Sakakibara, and Fumio Tokunaga

Key words. Retinal progenitor cells, Regeneration, Otx2

The neural retina is a good model for investigation of the regeneration of central nervous system (CNS), since it possesses clearly differentiated neurons and a well-organized structure. It is usually difficult to regenerate the damaged retina except for the embryonic retina of birds and frogs, but newts have this ability thorough their lifetime. After surgical removal of the neural retina in adult newts, the remaining retinal pigment epithelial cells proliferate with the loss of their pigment granules, and retain the retinal progenitor phenotype. This results in the reconstruction of a functional retina. To understand the molecular mechanism of the CNS regeneration, we tried to isolate the transcription factors required for normal development of vertebrate retinas. We screened a newt retinal cDNA library and isolated a cDNA similar to Otx2. The deduced amino acid sequence showed high identity with Otx2 of other vertebrates, human (87%), mouse (90%), Xenopus (87%), and zebrafish (88%). The expression pattern of the newt Otx2 is examined in adult normal retina, developing retina, and regenerating retina followed by the surgical removal of neural retina. Our in situ hybridization demonstrated that Otx2 is strongly expressed in all retinal progenitor cells from the earliest (mono-layered) stage to the several-layered stage in the regenerating retina. Similar expression patterns of Otx2 were observed in the developing retinas. Our results suggest that the molecular mechanisms of retinal development control the retinal regeneration from the earliest (monolayered) stage.

Graduate School of Science, Osaka University, 1-1 Machikaneyama, Toyonaka, Osaka 560-0043, Japan

Genes Involved in the Retinal Development Are Also Expressed in the Retinal Regeneration

Chikafumi Chiba, Kenta Nakamura, Yosuke Kikuchi, Kanako Susaki, and Takehiko Saito

Key words. Newt, Regeneration, Neurogenic genes

Newts can regenerate the functional retina from retinal pigment epithelial (RPE) cells. We examined here whether genes expressed in early retinal development are also expressed in this regeneration system. RNA samples were prepared from eye cups at selected days after surgical removal of the neural retina, and gene expressions during retinal regeneration were examined by reverse transcriptase-polymerase chain reaction. *Pax-6* genes were the so-called master control gene for eye development. Their expressions were detected around 10 days after surgery when RPE cells start cell division, and increased as regeneration progressed. Delta-Notch signaling negatively regulates the neuronal determination and contributes to keep neural stem cells in the developing retina. *Notch-1* was abruptly expressed, at the maximal level, around 20 days after surgery when one to two layers of retinal progenitor cells appeared. On the other hand, a Notch ligand and effector, *Delta-1* and *Hes-l*, were expressed at day 0 (i.e., normal eye cups without neural retina). *Delta-1* expression suddenly rose around 20 days after surgery and was sustained until at least 23 days. *Hes-1* expression slightly increased around 10 days and reached the maximum around 20 days. Expressions of these genes were down-regulated as the retina fully regenerated. *Neurogenin (Ngn)* is a positive regulator for neurogenesis. Expression of *Ngn* was detected at day 0. The expression level slightly increased around 10 days after surgery, reached the maximum around 23 days, and decreased as the retina fully regenerated. Correlating with the *Ngn* increase, neural markers such as *N-CAM* or voltage-gated Na^+ channel genes were expressed around 23 days. These observations may suggest that common molecular mechanisms are involved in retinal development and regeneration.

Institute of Biological Sciences, University of Tsukuba, 1-1-1 Tennodai, Tsukuba 305-8521, Japan

Difference in Survival and Axonal Regeneration Between Alpha and Beta Types of Cat Retinal Ganglion Cells

Takuji Kurimoto[1,2], Tomomitsu Miyoshi[1], Toru Yakura[1], Masami Watanabe[3], Osamu Mimura[2], and Yutaka Fukuda[1]

Key words. Cat, Retinal ganglion cells, Axonal regeneration, Survival, Apoptosis

Introduction

After optic nerve (ON) transection, most retinal ganglion cells (RGCs) undergo retrograde degeneration in adult mammals. When a piece of peripheral nerve (PN) is transplanted to the cut end of an ON, however, a small proportion of RGCs regenerate their axons along the graft [1, 2]. Furthermore, when these axons are reconnected with the target visual centers, the animals with the reconstructed visual pathway can regain some primitive visual functions such as pupillary light reflex and light–dark discrimination [3, 4].

For the recovery of visual function such as acute vision and shape recognition after ON damage, it is necessary to examine the survival ability and regeneration of RGCs in animals that have a well-differentiated visual system, such as the cat [5]. Cat RGCs can be classified into three types by morphological features: alpha cells, beta cells, and the rest [6]. Alpha cells have a large soma and dendritic field and physiologically corresponding Y cells have large receptive field centers and contribute to higher temporal sensitivity of vision. On the other hand, beta cells have a small dendritic field and physiologically corresponding X cells have small receptive field centers with brisk-tonic visual responses. In the area centralis (AC), beta cells are densely distributed and they are responsible for central vision such as visual acuity. The rest of RGCs, W cells in physiological terms, have the common characteristics of a small soma and a large dendritic field. They are further classified morphologically into gamma, delta, and epsilon cells, and so on [7].

[1] Department of Physiology and Biosignaling, Graduate School of Medicine, Osaka University, 2-2 Yamadaoka, Suita, Osaka 565-0871, Japan
[2] Department of Ophthalmology, Hyogo College of Medicine, 1-1 Mukogawa, Nishinomiya, Hyogo 663-8501, Japan
[3] Department of Physiology, Institute for Developmental Research, 713-8 Kamiya-cho, Kasugai, Aichi 480-0392, Japan

From our previous studies, it has been revealed that there are many differences in survival ability and axonal regeneration among these cell types. In the present review, we focus on the differences between alpha and beta cells.

Differences Between Alpha and Beta Cells in Regeneration Ability

Watanabe et al. [8] have previously reported that axotomized RGCs of adult cat can regenerate their axons into the PN-graft, like the RGCs in rodents. To assess the functional capacity of RGCs with regenerated axons, they examined whether or not the dendritic morphologies of RGCs with regenerated axons are preserved. The RGCs with regenerated axons revealed their dendritic morphology to be similar to that of intact cat RGCs even 2 months after PN transplantation [9]. Furthermore, RGCs with regenerated axons maintained their receptive field properties similar to those of intact cat RGCs [10]. These lines of evidence suggest a possibility that the RGCs with regenerated axons function as part of a parallel processing system when they are properly guided to the primary visual centers.

The regeneration ability was quite different between alpha and beta cells. In intact retinas, the relative proportion of alpha cells is 4.2% (Fig. 1A). In contrast, the relative proportion of alpha cells increased up to 23.9% in cat RGC populations with regenerated axons (Fig. 1B) [9]. This result was also confirmed with electrophysiological experiment (Fig. 1C). On the other hand, the relative proportion of beta cells with regenerated axons was almost equivalent to that of beta cells in intact retinas with both morphological and electrophysiological experiments (Fig. 1B,C). Therefore, alpha cells had a higher regeneration ability than beta cells. Furthermore, Watanabe et al. [11] compared the survival abilities of alpha and beta cells 2 months after ON transection (Fig. 1D). The relative proportion of surviving alpha cells increased to 16% in comparison to that of alpha cells in intact retinas. On the other hand, the

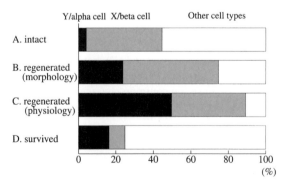

Fig. 1A–D. Relative proportions of each cell type of cat retinal ganglion cells (RGCs). **A** Intact cat RGCs. **B** RGCs with regenerated axons identified morphologically. **C** RGCs with regenerated axons identified electrophysiologically. **D** RGCs that survived 2 months after optic nerve (ON) transection. Note that the relative proportions of Y/alpha cells were much higher in both the surviving and regenerated RGC populations as compared with that in intact RGCs

Fig. 2A–C. Time course of the survival rates of alpha cells (**A, B**) and beta cells (**C**) within 2 weeks after ON transection. **A, C** Survival ratios of alpha and beta cells identified with Lucifer Yellow injections. **B** Ratios of the estimated total number of alpha cells in ON-transected retinas to those in the intact retinas. *Squares* and *vertical bars* in **A** and **C** indicate averages of normalized values of estimated cell number for the ON-transected retina to that in the intact retina with SD. *Solid lines* are regression lines calculated from the original values between days 3 and 14 (**A, C**), or days 3 and 7 and days 7 and 14 (**C**). Broken lines (**A, C**) draw the averages. Note that the survival rate of beta cells decreased rapidly from days 3 to 7, then slowly after day 7, whereas the survival rate of alpha cells decreased gradually. (From Wantanabe [12], with permission)

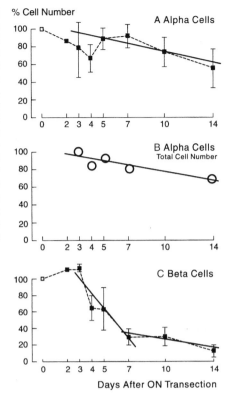

relative proportion of surviving beta cells markedly decreased to 9%. Thus, it is most probable that better regeneration ability of alpha cells can be ascribed to their better survival ability after ON transection.

Survival Time Course of Alpha and Beta Cells Within Two Weeks

To clarify the difference in the survival time course between alpha and beta cells after ON transection, Watanabe et al. [12] compared the survival time course of these cells within 2 weeks after ON transection (Fig. 2). The survival rate of axotomized alpha cells gradually decreased (the slope value = −3.03/day), and even 2 weeks after ON transection 64% of alpha cells survived (Fig. 2A; see also Fig. 2B). In contrast to alpha cells, the time course of surviving beta cells consisted of two phases: early phase (from days 3 to 7) and late phase (from days 7 to 14; Fig. 2C). In the early phase, the survival rate of axotomized beta cells rapidly decreased (the slope value = −18.6%/day), while the decrease was more gradual in the late phase (the slope value = −2.45%/day). Interestingly, the slope of the late phase of beta cells is close to the slope of alpha cells, suggesting that some common death mechanisms operate in alpha and beta cells. From the comparison of the survival time course between alpha and beta cells, it was concluded that the lower regeneration ability of beta cells is ascribable to the rapid death of axotomized beta cells shortly after ON transection.

Axotomized Beta Cells Rapidly Die by Apoptosis

From recent works done mostly on rats, it is now well established that some RGCs die by apoptosis after the ON injury [13]. After the ON transection, the number of rat RGCs that were stained with the TUNEL method peaked on day 6. Moreover, by intravitreal application of an inhibitor of caspase 3, which is an essential executor of apoptotic cell death, some axotomized RGCs can be rescued in adult rats [14]. Given these findings in rats, the possibility arose that in cat RGCs rapid death of beta cells after ON transection is due to apoptosis. We examined how axotomized RGCs undergo pyknosis, which is a hallmark of apoptotic morphological change, in Nissl-stained cat retinas up to 2 weeks after ON transection [15]. The proportion of pyknotic cells started to increase sharply from day 4 and reached its peak on day 6 after ON transection. Its time course corresponded very well to the early death phase of axotomized beta cells [12]. Furthermore, the proportion of pyknotic cells was the highest in the AC where beta cells were densely distributed. We further found that, after intravitreal injection of caspase 3 inhibitor, the survival of axotomized beta cells on day 7 after ON transection was significantly enhanced, whereas no such survival promoting effect was obtained in axotomized alpha cells. Taken together, these findings suggest that the rapid death of axotomized beta cells is mainly due to apoptosis, which is mediated by caspase 3.

Conclusion

From a series of studies on survival and regeneration of cat RGCs, we conclude that alpha cells have better survival and regeneration ability than do beta cells, probably due to the significant vulnerability of the latter to axotomy. It is possible that different intracellular mechanisms may underlie the survival and axonala regeneration of alpha and beta cells. The clarification of the difference will provide a new insight into the treatment of ophthalmic diseases such as glaucoma and Leber's optic neuritis.

References

1. So KF, Aguayo AJ (1985) Lengthy regrowth of cut axons from ganglion cells after peripheral nerve transplantation into the retina of adult rats. Brain Res 328:349–353
2. Vidal-Sanz M, Bray GM, Villegas-Pérez MP, et al. (1987) Axonal regeneration and synapse formation in the superior colliculus by retinal ganglion cells in the adult rat. J Neurosci 7:2894–2909
3. Whiteley SJO, Sauvé Y, Avilés-Trigueros M, et al. (1998) Extent and duration of recovered pupillary light reflex following retinal ganglion cell axon regeneration through peripheral nerve grafts directed to the pretectum in adult rats. Exp Neurol 154:560–572
4. Sasaki H, Inoue T, Iso H, et al. (1993) Light-dark discrimination after sciatic nerve transplantation to the sectioned optic nerve in adult hamsters. Vision Res 33:877–880
5. Watanabe M, Fukuda Y (2002) Survival and axonal regeneration of retinal ganglion cells in adult cats. Prog Retin Eye Res 21:529–553
6. Rodiek RW (1979) The visual pathway. Annu Rev Neurosci 2:193–225
7. O'Brien BJ, Isayama T, Richardson R, et al. (2001) Intrinsic physiological properties of cat retinal ganglion cells. J Physiol 538:787–802

8. Watanabe M, Sawai H, Fukuda Y (1991) Axonal regeneration of retinal ganglion cells in the cat geniculocortical pathway. Brain Res 560:330–333

9. Watanabe M, Sawai H, Fukuda Y (1993) Number, distribution, and morphology of retinal ganglion cells with axon regenerated into peripheral nerve graft in adult cats. J Neurosci 13:2105–2117

10. Miyoshi T, Watanabe M, Sawai H, et al. (1999) Receptive field properties of adult cat's retinal ganglion cells with regenerated axons. Exp Brain Res 124: 383–390

11. Watanabe M, Sawai H, Fukuda Y (1995) Number and dendritic morphology of retinal ganglion cells that survived after axotomy in adult cats. J Neurobiol 27:189–203

12. Watanabe M, Inukai N, Fukuda Y (2001). Survival of retinal ganglion cells after transection of optic nerve in adult cats: a quantitative study within two weeks. Vis Neurosci 18:137–145

13. Garcia-Valenzuela E, Gorczyca W, Darzynkiewicz Z, et al. (1994) Apoptosis in adult retinal ganglion cells after axotomy. J Neurobiol 25:431–438

14. Kermer P, Klöcker N, Labes M, et al. (1998) Inhibition of CPP-like proteases rescues axotomized retinal ganglion cells from secondary cell death in vivo. J Neurosci 18:4656–4662

15. Kurimoto T, Miyoshi T, Suzuki A, et al. (2003) Apoptotic death of beta cells after optic nerve transection in adult cats. J Neurosci (in press)

Survival-Promoting Effect of Electrical Stimulation on Axotomized Retinal Ganglion Cells

Tomomitsu Miyoshi[1], Takeshi Morimoto[1,2,3], Toru Yakura[1],
Yuka Okazaki[1], Takuji Kurimoto[1,4], Tetsu Inoue[1], Hajime Sawai[1],
Takashi Fujikado[2,3], Yasuo Tano[3], and Yutaka Fukuda[1]

Key words. Retinal ganglion cell, Axotomy, Electrical activity, Survival, Electrical stimulation

It has been well established that axotomized retinal ganglion cells (RGCs) of adult mammals can survive and regenerate their axons when the peripheral nerve is autografted [1]. However, the number of RGCs with regenerated axons is still limited. One major reason for poor regeneration is that many RGCs degenerate shortly after optic nerve (ON) transection. For example, half of the rat RGCs were lost within 1 week after intraorbital ON transection, and only 10% survived 2 weeks after axotomy [2]. The rescue of axotomized RGCs from such early retrograde degeneration is a prerequisite for the functional recovery of vision against ON damage.

Until now, there have been many attempts to enhance the survival of axotomized RGCs. A majority of the efforts apply various trophic factors and bioactive molecules into the vitreous of the damaged eye, for example, BDNF, NT-4/5, CNTF, FGF, IGF, GDNF, neurturin, macrophage inhibitory factor (MIF), TNF-α (for review, see [3]). In addition, various cells and tissues, such as peripheral nerve, Schwann cells, collicular proteoglycan, and an artificial graft with Schwann cells, and trophic factors have also been examined, and these attempts succeeded in the enhancement of RGC survival. The effect of these factors, however, became less marked as the day proceeded after the drug administration, and even prolonged administration of caspase inhibitor or NT-4/5 did not sustain survival promotion for a longer period [4, 5]. In addition to the simple application of trophic substances, several trials have been performed to transfer genes for long-term trophic support with a viral vector [6, 7] or cDNA electroporation [8]. A combination of TrkB gene transfer to RGCs and exogenous BDNF administration results in a marked enhancement of survival of axotomized RGCs within 2 weeks [9]. If exogenous BDNF is applied in the long term, this gene therapy

[1] Department of Physiology and Biosignaling, [2] Department of Visual Science, and [3] Department of Ophthalmology, Graduate School of Medicine, Osaka University, 2-2 Yamadaoka, Suita, Osaka 565-0871, Japan
[4] Department of Ophthalmology, Hyogo College of Medicine, 1-1 Mukogawa, Nishinomiya, Hyogo 663-8501, Japan

may promote the long-term survival of axotomized RGCs. From the viewpoint of clinical application of neuroprotective experiments, however, gene therapy has not been performed as the conventional treatment. Thus, different strategies from supplementing trophic factors are still desired as a neuroprotective treatment in the actual clinical situation.

It has been established that electrical activity of neurons controls cellular functions such as neurite extension, synaptic connectivity, and morphological change of dendrites, both in the developmental stage and in adulthood (for review, see [10]). Similarly, it is reasonable to suppose that electrical activity can also control the survival and regeneration of damaged neurons in the central nervous system.

Some results in our previous reports showed the relationship between electrical activity and RGC survival and/or axonal regeneration. One report was about degeneration of RGCs in the dark-reared cat after peripheral nerve transplantation [11]. After 2 months of rearing cats in a dark environment after surgery, there were fewer RGCs with regenerated axons than in the cats reared in a conventional light–dark environment. By intracellular dye injection, many RGCs in the dark-reared cat showed the morphological signs of degeneration such as nuclear condensation and vacuoles in both cytoplasm and dendrites. Moreover, some RGCs were unable to retain injected dye in the soma and lost it. More than half of the regenerated RGCs in dark-reared cats revealed such degenerative changes, whereas only a small proportion of RGCs showed such degenerative signs in cats reared in a conventional light–dark environment. These findings revealed that light stimulus is necessary for survival and axonal regeneration of RGCs. Survival enhancement by light exposure has also reported by Aigner et al. [12] on axotomized rat RGCs. Another result indicating the relationship between electrical activity and the survival of RGCs was the decrease of spontaneous discharge after ON transection [13]. The spontaneous discharge rate of cat RGCs became slightly low on day 5 after axotomy, and a majority of axotomized RGCs had no spontaneous activity on day 14. The fall of spontaneous activity in these RGCs may reflect the loss of synaptic inputs from bipolar or amacrine cells to RGCs and/or some changes in their membrane properties. From this observation, it can be thought that low or no spontaneous activity shortens the survival of axotomized RGCs.

Given these findings between the RGC survival and electrical activity, we hypothesized that electrical stimulation to the ON in vivo may rescue the axotomized RGCs from retrograde death. To verify this hypothesis, we examined whether the electrical stimulation to the transected ON improved the survival of rat RGCs 1 week after axotomy [14]. After the RGCs were retrogradely labeled with Fluorogold from the superior colliculi, the left ON was completely transected. Immediately after the ON transection, a train of electrical pulses of 20 Hz (a single pulse was monophasic, 50 μs in duration, and 50 μA in amplitude) was applied for 2 h to the ON stump via a pair of silver ball electrodes. The survival rates of RGCs 1 week after axotomy were evaluated by calculating the mean densities of RGCs from the counted number of retrogradely labeled RGCs in 12 areas of each treated and control retina. In normal retina, RGCs had fine-dotted Fluorogold in the perinuclear cytoplasm and proximal dendrites (Fig. 1a). Seven days after ON transection, the number of labeled RGCs declined markedly, and there was much debris of dead RGCs (Fig. 1b). In the retina with electrical stimulation, the surviving RGCs were obviously more numerous than were those in the retina with only axotomy (Fig. 1c).

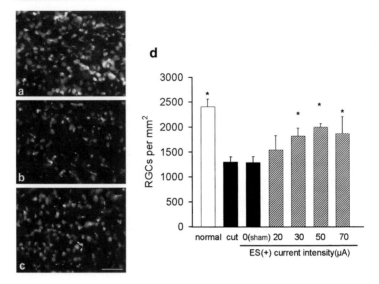

FIG. 1a–d. The effect of electrical stimulation on the survival of axotomized rat retinal ganglion cells (*RGCs*). The photomicrographs of Fluorogold-labeled RGCs in corresponding regions of an intact retina (**a**), 7 days after optic nerve (ON) transection without electrical stimulation (**b**), and with electrical stimulation of 50 μA (**c**). More RGCs survived in the retina with electrical stimulation than did those without electrical stimulation. *Bar* 50 μm. (**d**) The mean RGC densities of the groups with different current intensities of electrical stimulation. *Sham* indicates the group in which the electrodes were attached to the ON stump but no current was applied. Electrical stimulation over 30 μA significantly increased the densities of surviving RGCs. *Error bars* show standard deviations. Statistical analysis among groups was made using one-way ANOVA followed by Tukey test (one-way ANOVA; $p < 0.001$, *Tukey test; $p < 0.05$ compared to ON transection). Reprinted with permission from: Morimoto T, Miyoshi T, Fujikado T, et al. (2002) Neuroreport 13:227–230 Lippincott Williams & Wilkins

As a result, 83% of RGCs survived after electrical stimulation of the ON, whereas only 54% of RGCs survived without ON stimulation (Fig. 1d). The sham stimulation did not show any effect on the RGC survival rate. The survival-promoting effect of electrical stimulation depended on the intensity of electrical currents applied to the ON. From these results, the survival-promoting effect of electrical stimulation was clearly proved in the in vivo experiment. The enhancement of RGC survival by electrical stimulation of the ON was still clear 2 weeks after axotomy, though the absolute survival rate became lower in both treated and nontreated retinas. Though the mechanisms of survival promotion by electrical stimulation in vivo have not been elucidated yet, this electrical stimulation provides a great possibility for clinical application not only to axotomized RGCs but also to various neurological disorders.

References

1. So K-F, Aguayo AJ (1985) Lengthy regrowth of cut axons from ganglion cells after peripheral nerve transplantation into the retina of adult rats. Brain Res 328:349–354

2. Berkelaar M, Clarke DB, Wang Y-C, et al. (1994) Axotomy results in delayed death and apoptosis of retinal ganglion cells in adult rats. J Neurosci 14:4368–4374

3. Yip HK, So KF (2000) Axonal regeneration of retinal ganglion cells: effect of trophic factors. Prog Retin Eye Res 19:559–575

4. Clarke DB, Bray GM, Aguayo AJ (1998) Prolonged administration of NT-4/5 fails to rescue most axotomized retinal ganglion cells in adult rats. Vision Res 38:1517–1524

5. Kermer P, Klöcker N, Bähr M (1999) Long-term effect of inhibition of ced 3-like caspases on the survival of axotomized retinal ganglion cells in vivo. Exp Neurol 158:202–205

6. Di Polo A, Aigner LJ, Dunn RJ, et al. (1998) Prolonged delivery of brain-derived neurotrophic factor by adenovirus-infected Müller cells temporarily rescues injured retinal ganglion cells. Proc Natl Acad Sci USA 95:3978–3983

7. Schmeer C, Straten G, Kugler S, et al. (2002) Dose-dependent rescue of axotomized rat retinal ganglion cells by adenovirus-mediated expression of glial cell-line derived neurotrophic factor in vivo. Eur J Neurosci 15:637–643

8. Mo X, Yokoyama A, Oshitari T, et al. (2002) Rescue of axotomized retinal ganglion cells by BDNF gene electroporation in adult rats. Invest Ophthalmol Vis Sci 43:2401–2405

9. Cheng L, Sapieha P, Kittlerova P, et al. (2002) TrkB gene transfer protects retinal ganglion cells from axotomy-induced death in vivo. J Neurosci 22:3977–3986

10. Zhang LI, Poo MM (2001) Electrical activity and development of neural circuits. Nat Neurosci 4(Suppl):1207–1214

11. Watanabe M, Inukai N, Fukuda Y (1999) Environmental light enhances survival and axonal regeneration of axotomized retinal ganglion cells in adult cats. Exp Neurol 160:133–141

12. Aigner LJ, Shingel SA, Rasminsky M, et al. (1996) Light exposure enhances the survival of adult axotomized retinal ganglion cells. Abstr Soc Neurosci 22:305

13. Takao M, Miyoshi T, Watanabe M, et al. (2002) Changes in visual response properties of cat retinal ganglion cells within two weeks after axotomy. Exp Neurol 177:171–182

14. Morimoto T, Miyoshi T, Fujikado T, et al. (2002) Electrical stimulation enhances the survival of axotomized retinal ganglion cells in vivo. Neuroreport 13:227–230

A New Cl Channel Family Defined by Vitelliform Macular Dystrophy

Takashi Tsunenari[1,3], Hui Sun[2,3], King-Wai Yau[1,3], and Jeremy Nathans[1,2,3]

Key words. Bestrophin, Chloride channel, Retinal pigment epithelium, Channelopathy, Macular degeneration

Introduction

The macular region of the retina is responsible for high-acuity vision in the central part of the visual field. Macular degeneration leads to gradual decline and loss of the central vision, which we most heavily rely on in our everyday life. Most cases of macular degeneration are encountered as age-related macular degeneration, but their exact pathogenic mechanisms remain poorly defined. Although younger people develop macular degeneration much more rarely, some of them do develop the disease caused by mutated genes. A few such monogenic disorders have been identified and targeted for intensive study, both because of their intrinsic importance and because they may help understand the more common age-related forms of macular degeneration. These monogenic disorders include Stargardt disease, Sorsby's fundus dystrophy, dominant familial drusen, and vitelliform macular dystrophy (VMD, Best disease).

VMD, the subject of this article, is characterized by a distinctive accumulation of lipofuscin-like material within and beneath the retinal pigment epithelium (RPE). The VMD gene was isolated by positional cloning in 1998 [1, 2]. Its gene product, bestrophin, has at least 3 other homologues in the human genome, 4 in the *Drosophila* genome, and 24 in the *Caenorhabditis elegans* genome, but no function has yet been assigned to any members of this family, and they show no detectable homology to any protein of known function. There are, however, clues to the function of bestrophin: (i) VMD is associated with a reduction in the slow light peak of the electrooculogram, a component thought to reflect an increase in Cl conductance across the basolateral membrane of the RPE [3]; (ii) bestrophin has four predicted transmembrane domains; and (iii) among the disease-associated mutations reported, all are dominant

---

[1] Departments of Neuroscience, [2] Molecular Biology and Genetics, and [3] Howard Hughes Medical Institute, Johns Hopkins University School of Medicine, 725 North Wolfe Street, Baltimore, MD 21205, USA

and most of them cluster in or near the transmembrane domains [1–4]. All of these observations can potentially be explained if bestrophin is an oligomeric Cl channel. Recently, Marmorstein et al. [5] reported that bestrophin is localized to the basolateral membrane of the RPE, strengthening our speculation that bestrophin may be a Cl conductance. In our recent work, we tested our hypothesis by using a heterologous expression system and whole-cell recordings [6].

Membrane Currents Associated with Expressed Bestrophins

Human embryonic kidney (HEK) 293 cells transiently transfected with the cDNA coding for a bestrophin family member showed a whole-cell current not detected in control, mock-transfected cells. In our recording condition, interestingly, hBest1 (disease-associated bestrophin of human), hBest2 (second human homologue), ceBest1 (*C. elegans*), and dmBest1 (*Drosophila*) all had distinctive current-voltage (*I-V*) relations: very mild inward and outward rectifications for hBest1, an inward rectification for ceBest1, an outward rectification for dmBest1, and an *I-V* relation for hBest2 similar to hBest1 in negative voltages except for practically no rectification in positive voltages. Although its physiological significance remains unclear, this diversity suggests that the expressed protein constitutes part or all of the conducting channel. Otherwise, one would have to propose that each bestrophin homologue induces or unmasks a different type of endogenous ion channel. To confirm that expressed bestrophin constitutes an integral part of an ion channel, we examined whether mutations of bestrophin could change the properties of the expressed current. We found that, although wild-type hBest1 gave a whole-cell current that was irreversibly blocked by sulfhydryl-reactive reagents, a mutant with all five cysteine residues replaced by alanine produced a current resistant to the sulfhydryl-reactive reagents. This result strengthens the notion that bestrophin forms a part or all of the channel complex.

The bestrophin-associated channel shows properties of Cl channels. First, the bestrophin current is chloride-selective, in that the reversal potential of the hBest1 current was strongly dependent on extracellular Cl^- concentration, but not on extracellular Na^+ concentration. Bestrophin is also broadly permeable to other small inorganic anions. For hBest1, measurements of reversal-potential shifts in substitution experiments gave a permeability sequence $NO_3^- > I^- > Br^- > Cl^-$. For example, the permeability ratio P_{NO3}/P_{Cl} was 5.8 for hBest1 and 2.7 for hBest2. Finally, a Cl channel blocker, 4,4′-diisothiocyano-2,2′-stilbenedisulfonic acid (DIDS), reversibly blocked 90% of the hBest1 whole-cell current at 0.5 mM.

Oligomerization of Bestrophin

In many ion channels, identical or homologous subunits associate around an aqueous pore. Our coimmunoprecipitation experiments showed that bestrophin proteins do interact with each other. Although the bestrophin protein associates most efficiently with itself, it can associate with other members of the bestrophin family, albeit with lower efficiencies. For instance, the efficiency of association with hBest1 is hBest1 $\approx$ hBest2 >> ceBest1 $\approx$ dmBest1. The efficient association between hBest1 and hBest2 suggests that, like the subunits of many kinds of cation channels, different bestrophin subunits have the potential to heterooligomerize and produce combinatorial channel diversity. Very recently, Stöhr et al. [7] suggested that hBest2 (VMD2L1) is also expressed in RPE as hBest1, increasing the likelihood of this possibility. The channel properties of heterooligomerized bestrophins should be studied further.

To estimate the number of subunits in a channel complex, we performed an experiment based on the following reasoning. When Rim3F4 (R) -tagged and myc (M) -tagged hBest1s are coexpressed in the same cell and the channel complexes are immunoprecipitated by anti-M antibody, an ever-larger fraction of the channel complexes should contain only one M-tagged subunit with increasing the cDNA ratio of R-tagged hBest1. Thus, their total mole ratio of A/B should asymptotically approach (subunit number) minus 1. By immunopurifying [^{35}S]methionine-labeled hBest1 and measuring the radioactive band intensity by a phosphoimager, we observed an asymptotic ratio between 3:1 and 4:1, suggesting a stoichiometry of four or five subunits per complex.

Calcium Sensitivity of Bestrophin Cl Channel

Bestrophin was sensitive to intracellular Ca^{2+}. We loaded the photolyzable caged-Ca^{2+} compound, Ca^{2+}-NPEGTA, from a whole-cell pipette into an hBest1-transfected cell. The applied flash induced a significant current increase that was not observed in control experiments with mock-transfected cells. In our condition, the buffered free Ca^{2+} concentration was adjusted to $\approx 10\,nM$ before photolysis, but the Ca^{2+} concentration after uncaging of Ca^{2+} was not measured. Further studies with RPE cells and heterologous expression will be required to understand the Ca^{2+} action in more detail.

Vitelliform Macular Dystrophy-Associated Bestrophin Mutants

To investigate the mechanism by which hBest1 mutants cause VMD, we studied 15 disease-associated point mutants. Although the mutant proteins were similar to the wild-type protein in expression level in HEK cells, each mutant showed a considerably smaller or almost-completely eliminated bestrophin current. When the disease-associated mutant was cotransfected with the wild type, each of the tested mutants dominantly inhibited the conductance of wild-type hBest1. This dominant effect of the mutant protein is consistent with the dominant nature of Best disease. The dominance of the mutant alleles in causing disease is likely to result from the production of defective channels composed of both mutant and wild-type subunits.

Conclusion

Our results suggest that bestrophin functions as a Cl channel and VMD is induced by the malfunction of the bestrophin Cl channel in the RPE of the retina. A family of Cl channels defined by bestrophin has a putative secondary structure consisting of four transmembrane domains with N and C termini facing the cytosol [4]. If this structure prediction is correct, the bestrophin family shows a new architecture for the Cl channel. In view of this, it would be interesting to study the more detailed structure of the bestrophin channel and to understand the relation between its structure and function. Our findings also provide a direct explanation for the electrooculographic abnormalities in patients with VMD, and provide a framework for the pathogenesis of VMD. The photoreceptor outer segment depends almost entirely on the RPE for metabolic functions. The RPE is highly active in the transport of various metabolites such as protons, cations, bicarbonate, organic anions, and various other molecules, which are coupled directly or indirectly to chloride movement [8]. Thus, defects in chloride homeostasis might create an imbalance in pH and/or ionic environments. Given the enormous flux of material handled by the RPE, including phagocytosed photoreceptor outer segments, nutrients, and metabolites, such an imbalance may result in impaired transport and accumulation of debris.

Acknowledgments. The work presented here was supported by the Howard Hughes Medical Institutes (J.N. and K.-W.Y.), the Foundation Fighting Blindness (J.N.), the National Institutes of Health (K.-W.Y.), and a Long-Term Fellowship from the Human Frontier Science Program (T.T.).

References

1. Marquardt A, Stöhr H, Passmore LA, et al. (1998) Mutations in a novel gene, *VMD2*, encoding a protein of unknown properties cause juvenile-onset vitelliform macular dystrophy (Best's disease). Hum Mol Genet 7:1517–1525
2. Petrukhin K, Koisti MJ, Bakall B, et al. (1998) Identification of the gene responsible for Best macular dystrophy. Nat Genet 19:241–247
3. Gallemore RP, Maruiwa F, Marmor MF (1998) Clinical electrophysiology of the retinal pigment epithelium. In: Marmor MF, Wolfensberger TJ (eds) The retinal pigment epithelium. Oxford University Press, Oxford, pp 199–223
4. Bakall B, Marknell T, Ingvast S, et al. (1999) The mutation spectrum of the bestrophin protein – functional implications. Hum Genet 104:383–389
5. Marmorstein AD, Marmorstein LY, Rayborn M, et al. (2000) Bestrophin, the product of the Best vitelliform macular dystrophy gene (*VMD2*), localizes to the basolateral plasma membrane of the retinal pigment epithelium. Proc Natl Acad Sci USA 97:12758–12763
6. Sun H, Tsunenari T, Yau K-W, et al. (2002) The vitelliform macular dystrophy protein defines a new family of chloride channels. Proc Natl Acad Sci USA 99:4008–4013
7. Stöhr H, Marquardt A, Nanda I, et al. (2002) Three novel human VMD2-like genes are members of the evolutionary highly conserved RFP-TM family. Eur J Hum Genet 10:281–284
8. Hughes BA, Gallemore RP, Miller SS (1998) Transport mechanisms in the retinal pigment epithelium. In: Marmor MF, Wolfensberger TJ (eds) The retinal pigment epithelium. Oxford University Press, Oxford, pp 103–134

Pharmacological Aspects of Nilvadipine-Induced Preservation of Retinal Degeneration in RCS Rat Analyzed by mRNA Profiling Assay

Futoshi Ishikawa, Hiroshi Ohguro, Ikuyo Maruyama, Yoshiko Takano, Hitoshi Yamazaki, Tomomi Metoki, Yasuhiro Miyagawa, Motoya Sato, Kazuhisa Mamiya, and Mitsuru Nakazawa

Summary. In our study, we found that the Ca^{2+} antagonist nilvadipine was beneficial for the preservation of photoreceptor cells in The Royal College of Surgeons (RCS) rats. Here, in order to elucidate mechanisms of its nilvadipine-induced photoreceptor preservation, we analyzed altered gene expressions of the retina in RCS rats administered nilvadipine by mRNA profiling assay. Total RNA isolated from the retina with or without nilvadipine was converted into cDNA. Utilizing DNA microarray analysis methods, we compared the overall expression patterns for 1101 genes that were commonly expressed in rodent. Of the total genes, the expression of less than 30 genes was altered significantly including that of several genes involved in cellular regulation. On the basis of these data, it is suggested that microarray analysis is a useful tool and applicable for studying the pharmacological effects of several drugs including Ca^{2+} channel blockers to retinal degeneration.

Key words. Nilvadipine, RCS rat, mRNA profiling assay, Apoptosis retinitis pigmentosa

Introduction

The Royal College of Surgeons (RCS) rat, in which the retinal pigment epithelium (RPE) cell is affected by the retinal dystrophy (rdy)- mutation and continuously expresses the rdy- phenotype [1], is the best characterized animal model for the study of human retinitis pigmentosa (RP). It was suggested that an inability of phagocytosis of the shed tips of rod outer segment (ROS) debris by RCS RPE is primarily involved as a possible mechanism causing the retinal degeneration [2]. D'Cruz et al. [3] have found a small deletion of RCS DNA that disrupts the gene encoding the receptor tyrosine kinase Mertk as the rdy locus of the RCS rat. Interestingly, the identical mutations in *Mertk* have recently been identified in human RP patients [4]. Therefore, these observations suggest that if some therapy may preserve retinal degeneration in RCS rat, such a therapy should be beneficial and applicable to human RP. In our recent study, we have found that systemic administration of nilvadipine, which

Department of Ophthalmology, Hirosaki University School of Medicine, 5 Zaifucho, Hirosaki 036-8562, Japan

is the Ca^{2+} antagonist that most effectively penetrates the central nervous system, caused significant preservation of retinal morphology and functions of electroretinogram responses in RCS rats during the initial stage of the retinal degeneration. In addition, studies using immunohistochemistry, reverse transcriptase-polymerase chain reaction, and Western blotting revealed significant enhancement of rhodopsin kinase and α-A-crystallin expressions, which are specifically reduced in RCS rat retina as compared with control rats, and suppression of caspase 1 and 2 expressions in the retina of nilvadipine-treated rats [5]. However, the molecular mechanisms by which systemically administered nilvadipine prevent the RCS retinal degeneration remain as yet undefined.

In the present study, in order to elucidate what kinds of mechanisms are involved in the nilvadipine-dependent preservation of retinal morphology and function in RCS retinal degeneration, an mRNA profiling assay was performed.

Materials and Methods

All experimental procedures were designed to conform to both the ARVO statement for Use of Animals in Ophthalmic and Vision Research and our own institution's guidelines. Unless otherwise stated, all procedures were performed at 4°C or on ice using ice-cold solutions. Nilvadipine was a generous gift from Fujisawa Pharmaceutical (Tokyo, Japan). Anti-FGF2 antibody and anti-Arc antibody were purchased from Santa Cruz Biotechnology (San Francisco, CA). The specificity and titers of all antibodies were examined by Western blot and enzyme-linked immunosorbant assay using spraque-Dawley (SD) rat retinal soluble fractions.

Anesthesia of the Animals

In the present study, 3- to 5-week-old inbred RCS (*rdy-/-*) rats (Crea, Tokyo, Japan) reared in cyclic light conditions (12 h on/ 12 h off) were used. For anesthesia induction, rats inhaled diethyl ether. Once unconscious, the animals were injected intramuscularly with a mixture of ketamine (80–125 mg/kg) and xylazine (9–12 mg/kg). Adequacy of the anesthesia was tested by tail clamping, and supplemental doses of the mixture were administered intramuscularly if needed.

Drug Administration

Nilvadipine was dissolved in a mixture of ethanol:polyethylene glycol 400:distilled water (2:1:7) at a concentration of 0.1 mg/ml, diluted twice with physiological saline before use, and injected intraperitoneally (1.0 ml/kg) into anesthetized rats every day early in the morning for 2 weeks. In control rats, the same solution without nilvadipine (vehicle solution) was administered as described earlier.

DNA Microarray Analysis

Total RNA from retinas was isolated using ISOGEN reagent according to the procedure recommended by the manufacture (Nippon Gene, Tokyo, Japan). Three micrograms of total RNA was reverse transcribed in the presence of 300 units of SuperScript

II RNase H Reverse Transcriptase (Life Technologies, Gaitherburg, MD), 100 μCi of [^{33}P]deoxycytidine triphosphate (10 mCi/ml, 3000 Ci/mmol; ICN, Costa Mesa, CA), and 2 μg of oligo(dT) (10-20-mer; Invitrogen, Carlsbad, CA). Reactions were each carried out at 37°C for 90 min in 30 μl of buffer consisting of 50 mM TrisHCl (pH 8.3), 75 mM KCl, 3 mM MgCl2, 3.3 mM dithiothreitol, and 1 mM each of deoxyadenosine triphosphate, deoxyguanosine triphosphate, and deoxythymidine triphosphate. The resulting ^{33}P-cDNA probes were purified with Bio-Spin 6 Chromatography Columns (Bio-Rad, Hercules, CA) following the manufacturer's instructions. Human Named Genes GeneFilters (GF211 from Invitrogen) were used for differential expression screening. The membranes were first pretreated with boiled 0.5% sodium dodecyl sulfate (SDS) for 10 min. Prehybridizations were performed for 6 h at 42°C in Micro-Hyb hybridization solution (Invitrogen) with poly(dA) and denatured Cot-1 as blocking reagents. The column-purified and denatured probes were then added and hybridized at 42°C for 16 h. After hybridization, the membranes were washed twice in a solution containing 2× standard saline citrate (SSC) and 1% SDS for 20 min at 50°C followed by five additional washes performed at the same temperature in a washing solution consisting of 0.5× SSC and 1% SDS for 15 min each. The membranes were then exposed to phosphor image screens for 16 h. Images were acquired by using a Cyclone Phosphor System (Packard, Palo Alto, CA), and analyzed by using the PATH-WAYS 2.1 (Invitrogen) and EXCEL (Microsoft, Redmand, WA). GeneFilters from the same manufacturing lot were used in all three separate experiments.

The hybridization for each sample at each time point was done in triplicate using three separate GeneFilters. Normalized intensities were calculated from each Gene-Filter by first subtracting a constant background value as reported by PATHWAYS and dividing each point by the average intensity for that GeneFilter and multiplying the result by 2000. Gene expression ratios from each experiment were calculated by using the average normalized intensities from each of the GeneFilters for a specific RNA sample (shear stressed to control values). The expression ratios reported are the average and standard deviations from the three separate experiments. Only genes with average normalized intensities of 100 or above were studied.

Genes with a ratio of 2.0 or above were considered positively regulated, whereas those that had a ratio of 0.5 or below were considered negatively regulated. The plate was scanned and the fold changes in RNA levels were quantitated by normalizing relative to glyceraldehyde-3-phosphate dehydrogenase (GAPDH).

Results and Discussion

Recently developed microarray analysis is a very powerful tool for understanding complicated physiological, pathological, and pharmacological phenomena [6]. By utilizing this methodology in the present study, we found that fewer than 30 out of 1101 genes, which are involved in a variety of cellular regulatory mechanisms, were up-regulated or down-regulated in RCS rat retina after systemic administration of nilvadipine, which has recently been identified as causing protective effects against RCS retinal degeneration.

Ca^{2+} antagonists, which have been widely used as treatments for systemic hypertension, inhibit the entry of calcium ion intracellularly, relax vascular smooth muscle cells, and increase regional blood flow in several organs [7]. As major Ca^{2+} antagonists, a dihydropyridine (DHP) derivative such as nifedipine, nicardipine, and nilvadipine; a benzothiazepine derivative such as diltiazem; and a phenylalkylamine derivative such as verapamil are known to be used in our clinical practices [8]. It was shown that these drugs have different properties in the specificity and kinetics of blocking of Ca^{2+} channels. On the basis of kinetics and voltage-dependent properties, the Ca^{2+} channels have been classified into two groups: one is activated by small depolarizations and is then subsequently inactivated, and is called the low-voltage-activated (LVA) Ca^{2+} channel, whereas the other is activated by larger depolarizations and shows little inactivation, and is called the high-voltage-activated (HVA) Ca^{2+} channel. In terms of retinal Ca^{2+} channels, it was revealed that L-type HVA Ca^{2+} currents of the channels in the vertebrate photoreceptors were recognized. Recently, a novel Ca^{2+} channel gene, *CACN1F*, encoding α_{1F}, a retina-specific α_1 subunit of L-type HVA Ca^{2+} channels, in which mutations in *CACN1F* cause incomplete X-linked congenital stationary night blindness (CSNB2), was identified [9, 10]. Taken together with the fact that most DHPs and diltiazem are L-type HVA Ca^{2+} channel blockers [8], we can reasonably speculate that these drugs may react with retinal L-type HVA Ca^{2+} channels, and may presumably prevent photoreceptor cell death. Therefore, it seems very critical to understand the pharmacological mechanisms of nilvadipine effects upon degenerating retinas at molecular levels in preparation for its clinical application in the near future. Therefore, further study to identify the up-regulated or down-regulated genes in the retina upon administration of nilvadipine is our next project.

References

1. Mullen RJ, LaVail MM (1976) Inherited retinal dystrophy: primary defect in pigment epithelium determined with experimental rat chimeras. Science 201:1023–1025
2. Edwards RB, Szamier RB (1977) Defective phagocytosis of isolated rod outer segments by RCS rat retinal pigment epithelium in culture. Science 197:1001–1003
3. D'Cruz PM, Yasumura D, Weir J, et al. (2000) Mutation of the receptor tyrosine kinase gene *Mertk* in the retinal dystrophic RCS rat. Hum Mol Genet 9:645–651
4. Gal A, Thompson DA, Weir J, et al. (2000) Mutations in *MERTK*, the human orthologue of the RCS rat retinal dystrophy gene, cause retinitis pigmentosa. Nat Genet 26:270–271
5. Yamazaki H, Ohguro H, Yasumura D, et al. (2002) Preservation of retinal morphology and functions in royal college surgeons rat by nilvadipine, a Ca(2+) antagonist. Invest Ophthalmol Vis Sci 43: 919–926
6. Mukai T, Joh K, Arai Y, et al. (1986) Tissue-specific expression of rat aldolase A mRNAs. Three molecular species differing only in the 5′-terminal sequences. J Biol Chem 261:3347–3354
7. Imai J, Omata K, Kamizuki M, et al. (1996) Antihypertensive effects of nilvadipine determined by ambulatory blood pressure monitoring. Ther Res 17:1847–1856
8. Triggle DH (1981) In: Weiss GB (ed) New perspectives on calcium antagonists. American Physiological Society, Bethesda, pp 1–21
9. Bech-Hansen NT, Naylor MJ, Maybaum TA, et al. (1998) Loss-of-function mutations in a calcium-channel alpha1-subunit gene in Xp11.23 cause incomplete X-linked congenital stationary night blindness. Nat Genet 19:264–267
10. Strom TM, Nyakatura G, Apfelstedt-Sylla E, et al. (1998) An L-type calcium-channel gene mutated in incomplete X-linked congenital stationary night blindness. Nat Genet 19:260–263

Nilvadipine, a Ca^{2+} Antagonist, Effectively Preserves Photoreceptor Functions in Royal College of Surgeons Rat

Hitoshi Yamazaki[1], Hiroshi Ohguro[1], Ikuyo Maruyama[1],
Yoshiko Takano[1], Tomomi Metoki[1], Futoshi Ishikawa[1],
Yasuhiro Miyagawa[1], Kazuhisa Mamiya[1], Mitsuru Nakazawa[1],
Hajime Sawada[2], and Mari Dezawa[2]

Summary. The Royal College of Surgeons (RCS) rat is the most extensively studied animal model for understanding the molecular pathology of inherited retinal degeneration, such as retinitis pigmentosa (RP). We recently found lower levels of rhodopsin kinase expression in RCS rats as compared with control rats, leading us to speculate that misregulation of the phototransduction pathways by the low levels of rhodopsin phosphorylation might be a critical step causing the retinal degeneration. Here, effects of suppression of recoverin-dependent inhibition of rhodopsin phosphorylation on the retinal degeneration in RCS rats by lowering intracellular Ca^{2+} levels by intraperitoneal administration of nilvadipine, a calcium antagonist, and retinal functions and morphology were analyzed. We found that systemic administration of nilvadipine caused remarkable preservation of photoreceptor functions, electroretinogram responses, and retinal morphology in RCS rats during the initial stage of the retinal degeneration. On the basis of these data, it was strongly suggested that nilvadipine is beneficial for the preservation of photoreceptor cells in RCS rats and can potentially be used to treat some RP patients.

Key words. Nilvadipine, Ca^{2+} antagonist, RCS rat, Retinal degeneration, Rhodopsin kinase

Introduction

The Royal College of Surgeons (RCS) rat, in which the retinal pigment epithelium (RPE) cell is affected by the retinal dystrophy (rdy)- mutation and continuously expresses the rdy- phenotype [1], has been the most widely used animal model for the study of retinitis pigmentosa (RP). In terms of the molecular pathology of the retinal degeneration, it was suggested that an inability of phagocytosis of the shed tips of rod outer segment (ROS) debris by RCS RPE is primarily involved [2]. Recently, D'Cruz et al. [3] have used a positional cloning approach to study the rdy locus of the

[1] Department of Ophthalmology, Hirosaki University School of Medicine, 5 Zaifucho, Hirosaki 036-8562, Japan
[2] Department of Anatomy, Yokohama City University School of Medicine, 3-9 Fukuura, Kanazawa-ku, Yokohama 236-0004, Japan

RCS rat, and they discovered a small deletion of RCS DNA that disrupts the gene encoding the receptor tyrosine kinase Mertk, which may be a molecular target for ingestion of outer segments by RPE cells. In contrast, RCS photoreceptor cells are considered to be normal in their structure and functions since RCS photoreceptor cells can survive after retinal RPE transplantation [4]. However, it was found that several changes occurred in RCS ROS including opsin [5], arrestin [6], and ROS protein phosphorylation levels [7], which might affect quenching of the phototransduction pathway in RCS. In fact, we recently found significantly lower levels of mRNA expressions of α-A crystalline and rhodopsin kinase (RK), which are thought to be involved in post-Golgi processing of opsin and rhodopsin phosphorylation, respectively, in RCS rats than in the control rats [8]. Therefore, based upon these observations, we suggested that low levels of rhodopsin phosphorylation might cause misregulation of the phototransduction pathways in rod photoreceptor cells, resulting in their degeneration. This idea is supported by experimental evidence that absence of rhodopsin phosphorylation in transgenic mice carrying rhodopsin mutations caused retinal degeneration [9]. Since recoverin, a retina-specific Ca^{2+}-binding protein, negatively regulates rhodopsin phosphorylation by rhodopsin kinase in a Ca^{2+}-dependent manner [10], it is plausible that suppression of recoverin-dependent inhibition of rhodopsin kinase by the lowering of intracellular Ca^{2+} levels by some drugs may be efficient in the preservation of photoreceptor cells in RCS rats. Interestingly, Frasson et al. [11] recently reported rod photoreceptor rescue by D-*cis*-diltiazem, a Ca^{2+} channel blocker in a different animal model of RP, rd mouse, in which the gene encoding cyclic guanosine monophosphate phosphodiesterase is affected. In consideration of these data, we assume that the regulation of intracellular Ca^{2+} levels may be a possible therapy to prevent progressive retinal degeneration in RP and its animal models.

To test our hypothesis, nilvadipine, the most effective penetrator of the central nervous system among Ca^{2+} antagonists used in clinical practice [12], was administered to 3-week-old RCS rats, which is the age when the degenerative changes in photoreceptor cells are known to start.

Then the retinal function was evaluated by electroretinography (ERG) and a histological study was performed.

Materials and Methods

All experimental procedures were designed to conform to both the ARVO statement for Use of Animals in Ophthalmic and Vision Research and our own institution's guidelines.

Anesthesia of the Animals

In the present study, 3- to 5-week-old inbred RCS (*rdy-/-*) rats (Crea, Tokyo, Japan) reared in cyclic light conditions (12h on/ 12h off) were used. For anesthesia induction, rats inhaled diethyl ether. Once unconscious, the animals were injected intramuscularly with a mixture of ketamine (80–125mg/kg) and xylazine (9–12mg/kg).

Drug Administration

Nilvadipine (Fujisawa Pharmaceutical, Tokyo, Japan) was dissolved in a mixture of ethanol:polyethylene glycol 400:distilled water (2:1:7) at a concentration of 0.1 mg/ml, diluted twice with physiological saline before use, and injected intraperitoneally (2.0 ml/kg) into anesthetized rats every day early in the morning for 2 weeks. In control rats, the same solution without nilvadipine (vehicle solution) was administered.

Light Microscopy

Anesthetized animals were transcardially perfused with a total 100 ml of 82 mM sodium phosphate buffer, pH 7.2, containing 4% paraformaldehyde. Enucleated eyes were embedded in paraffin and sectioned at a 3-μm thickness, mounted on subbed slides, and dried. The sections were processed with hematoxylin-eosin staining after deparaffinization with graded ethanol and xylene solutions.

Electroretinography

The anesthetized animals were kept in dark adaptation for at least 2 h in an electrically shielded room. The pupils were dilated with drops of 0.5% tropicamide. The scotopic ERG response was recorded as described previously [11].

Statistical Analysis

The data are shown as mean SD. P values <0.05 were considered significantly different as assessed by the Mann-Whitney test.

Results and Discussion

Recently, it has been reported that lowering of intracellular Ca^{2+} by Ca^{2+} antagonist and other drugs effectively suppresses the retinal neuronal apoptosis induced in some experimental animal models by ischemia reperfusion [13] and intravitreal injection of N-methyl-D-aspartate [14]. These observations allowed us to speculate that similar effects by Ca^{2+} antagonist may be expected in inherited retinal degeneration such as RP.

In the present study, to study the effects of nilvadipine, which is the most effective penetrator of the central nervous system among Ca^{2+} antagonists used in clinical practice [12], on the retinal morphology of RCS rat, nilvadipine or its vehicle solution as a control was systemically administered to 3-week-old RCS rats every day for 2 weeks, and the thickness of each retinal layer was compared with each other at 4 and 5 weeks of age. As shown in Fig. 1, all retinal layers, and the inner plexiform layer (IPL), inner nuclear layer (INL) and outer plexiform layer (ONL) were significantly thicker in 4- and 5 week-old RCS rats administered nilvadipine, respectively, as compared with their controls. This suggests that nilvadipine has a significant effect for preservation against thinning of retinal layers of RCS rat during its retinal degeneration.

FIG. 1. Thickness of retinal layers. Hematoxylin-eosin staining of retinal sections near the posterior pole from Royal College of Surgeons (RCS) rat eyes at the age of 4 weeks and 5 weeks treated with nilvadipine (2, 4) or its vehicle solution (1, 3). Photographs of the sections were taken, and each retinal layer was measured. GCL, ganglion cell layer; IPL, inner plexiform layer; INL, inner nuclear layer; OPL, outer plexiform layer; ONL, outer nuclear layer; OS, outer segment. P < 0.01 (Mann-Whitney test)

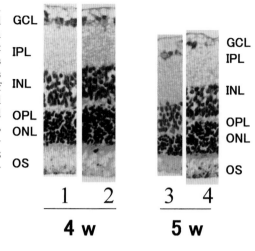

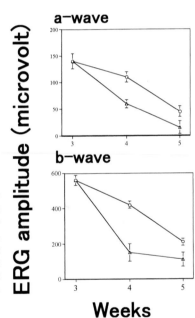

FIG. 2. Changes in scotopic electroretinography (ERG) in RCS rats treated with nilvadipine. RCS rats were treated with nilvadipine (open circles), or vehicle solution for nilvadipine (open triangles), respectively, every day from 3 weeks old. During the 2 weeks after the operation, scotopic ERG was recorded once a week. ERG measurements were performed in eight eyes in each condition and a-wave and b-wave amplitudes were plotted. P < 0.01 (Mann-Whitney test)

Next we also investigated drug effects of nilvadipine on retinal function by ERG. As shown in Fig. 2, nilvadipine-treated retina showed a significant preservation of a- and b-wave amplitudes as compared with the control at 4 weeks (P < 0.001). The preservation effect on ERG response by nilvadipine was less, but still significant (P < 0.01), at 5 weeks. Therefore, taken together, these data strongly suggested that administration of nilvadipine produces significant preservation of retinal morphology and function.

Conclusion

In conclusion, our present study strongly suggested that nilvadipine may be effective in the preservation of photoreceptor cells during retinal degeneration in RCS rats and this may lead to the clinical application of nilvadipine in the treatment of RP.

References

1. Mullen RJ, LaVail MM (1976) Inherited retinal dystrophy: primary defect in pigment epithelium determined with experimental rat chimeras. Science 201:1023–1025
2. Edwards RB, Szamier RB (1977) Defective phagocytosis of isolated rod outer segments by RCS rat retinal pigment epithelium in culture. Science 197:1001–1003
3. D'Cruz PM, Yasumura D, Weir J, et al. (2000) Mutation of the receptor tyrosine kinase gene *Mertk* in the retinal dystrophic RCS rat. Hum Mol Genet 9:645–651
4. Faktrovich EG, Steinberg RH, Yasumura D, et al. (1990) Photoreceptor degeneration in inherited retinal dystrophy delayed by basic fibroblast growth factor. Nature 347:83–86
5. Nir I, Sagie G, Papermaster DS (1987) Opsin accumulation in photoreceptor inner segment plasma membranes of dystrophic RCS rats. Invest Ophthalmol Vis Sci 28:62–69
6. Mirshahi M, Thillaye B, Tarraf M, et al. (1994) Light-induced changes in S-antigen (arrestin) localization in retinal photoreceptors: differences between rods and cones and defective process in RCS rat retinal dystrophy. Eur J Cell Biol 63:61–67
7. Heth CA, Schmidt SY (1992) Protein phosphorylation in retinal pigment epithelium of Long-Evans and Royal College of Surgeons rats. Invest Ophthalmol Vis Sci 33: 2839–2847
8. Maeda A, Ohguro H, Maeda T, et al. (1999) Low expression of αA-crystallin and rhodopsin kinase of photoreceptors in retinal dystrophy rat. Invest Ophthalmol Vis Sci 40:2788–2794
9. Chen J, Makino CL, Peachery NS, et al. (1995) Mechanisms of rhodopsin inactivation in vivo as revealed by a COOH-terminal truncation mutant. Science 267:374–377
10. Kawamura S (1991) Rhodopsin phosphorylation as a mechanism of cyclic GMP phosphodiesterase regulation by S-modulin. Nature 362:855–857
11. Frasson M, Sahel JA, Fabre M, et al. (1999) Retinitis pigmentosa: rod photoreceptor rescue by a calcium-channel blocker in the rd mouse. Nat Med 5:1183–1187
12. Ohtsuka M, Yokota M, Kodama I, et al. (1989) New generation dihydropyridine calcium entry blockers: in search of greater selectivity for one tissue subtype. Gen Pharmacol 20:539–556
13. Takahashi K, Lam TT, Edward DP, et al. (1992) Protective effects of flunarizine on ischemic injury in the rat retina. Arch Ophthalmol 110:862–870
14. Osborne NN, DeSantis L, Bae JH, et al. (1999) Topically applied betaxolol attenuates NMDA-induced toxicity to ganglion cells and the effects of ischaemia to the retina. Exp Eye Res 69:331–342
15. Bush RA, Konenen L, Machida S, et al. (2000) The effect of calcium channel blocker D-*cis*-diltiazem on photoreceptor degeneration in the rhodopsin Pro23His rat. Invest Ophthalmol Vis Sci 41:2697–2701

Aberrantly Expressed Recoverin in Tumor Tissues from Gastric Cancer Patients

YASUHIRO MIYAGAWA, HIROSHI OHGURO, IKUYO MARUYAMA,
YOSHIKO TAKANO, HITOSHI YAMAZAKI, FUTOSHI ISHIKAWA,
TOMOMI METOKI, KAZUHISA MAMIYA, and MITSURU NAKAZAWA

Key words. Cancer-associated retinopathy (CAR), Recoverin, Proliferation, Paraneoplastic syndrome

Introduction

A variety of neurological symptoms characterizing paraneoplastic syndromes, such
as Lambert-Eaton myasthenic syndrome and paraneoplastic cerebellar degeneration,
are well known to be associated with malignant neoplasms even though there is no
evidence of nervous system invasion [1]. The possible pathogenesis is an autoimmune
reaction toward aberrantly expressed neuron-specific antigens in cancer cells that
presumably triggers immunological responses and causes neuronal cell degeneration.
Cancer-associated retinopathy (CAR) has been identified as an ocular manifestation
of paraneoplastic syndrome [2]. CAR is frequently found in patients with small cell
carcinoma of the lung and other malignancies, and is clinically characterized with
retinitis pigmentosa-like retinal degeneration including photopsia, progressive visual
loss with a ring scotoma, attenuated retinal arterioles, and abnormalities of the a- and
b-waves of an electroretinogram (ERG) [3]. Histopathology has shown primarily a
loss of photoreceptor cells [2]. In terms of CAR autoantigen, a photoreceptor-specific,
26-kDa calcium-binding protein called recoverin [4] and other retinal antigens were
identified. Recoverin is known to play an important role in light and dark adaptation
by regulating rhodopsin phosphorylation and dephosphorylation in a calcium-
dependent manner [5]. Alternatively, recoverin is also known to be a highly immuno-
genic molecule and to experimentally cause uveoretinitis in rat by its immunization
[6]. A mechanism that has been postulated for the apoptotic cell death of photoreceptor cells observed in CAR is that antirecoverin antibody generated by unknown
mechanisms in some cancer patients penetrates into the photoreceptor cells via the
peripheral circulation and causes misregulation of the phototransduction pathway by
blocking recoverin function [7]. Several studies had suggested that, in the initial step
of CAR pathogenesis, aberrant expression of retina-specific recoverin, serving as

Department of Ophthalmology, Hirosaki University School of Medicine, 5 Zaifucho, Hirosaki
036-8562, Japan

shared antigen between tumor cells and retina, was critical since such aberrant expression of recoverin had been identified in tumor cells in CAR patients but not in non-CAR cancerous patients [8]. However, in contrast, we have recently found that aberrant expression of retinal-specific recoverin occurred in more than 50% of several kinds of cancer cells and their cell lines obtained from non-CAR cancerous patients [7, 9]. Furthermore, we have thus far made the following experimental observations: (1) immunofluorescence labeling by affinity-purified recoverin antibody revealed the immunoreactivity toward recoverin as a granular pattern within the cancer cells [9]; (2) transfection with human recoverin cDNA to A549 cells from lung adenocarcinoma, which did not express recoverin, exhibited a significant reduction in cell proliferation [9]; (3) significant changes were observed in protein phosphorylation patterns in the cytosol of A549 cells upon addition of purified recoverin in a calcium-dependent manner [7]; and (4) recoverin-specific cytotoxic T lymphocytes (CTL) exist in the peripheral circulation of CAR patients, and these CTL can recognize tumor expressing recoverin [10]. These observations correspond well with the favorable prognosis characteristic of malignant tumors underlying CAR. Therefore, it is very important that such recoverin expression is indeed present in tumor tissues obtained from cancerous patients, as observed in cancer cell lines.

In the present study, to test this, we performed immunocytochemistry using specific antibody toward human recoverin and cancer tissues obtained surgically from patients with gastric cancer.

Materials and Methods

The studies were performed in accordance with our institution's guidelines and the Declaration of Helsinki on Biomedical Research Involving Human Subjects, and protocols were approved by the institution's Committee for the Protection on Human Subjects. Informed consents were obtained from all patients.

Patients

Three gastric carcinoma tissues were obtained from three patients who underwent surgery at Hirosaki University Hospital. The definitions of stage grouping were made according to the fifth edition of the tumor-node-metastasis classification for gastric cancer by Union Internationale Contre le Cancer (UICC-TNM) [11].

Preparation of Antihuman Recoverin Antibody

A pET-3d vector-inserted human recoverin was generously provided from Professor Satoru Kawamura, Osaka University, Japan. Expression of human recoverin was induced with 0.1 mM isopropyl-β-D(−)-thiogalactopyranoside for 3 h at 37°C. After centrifugation at 4500 rpm for 10 min, the cell pellet was suspended in 10 ml of phosphate buffered saline (PBS) containing 0.5% Triton-X 100 and sonicated for 3 min on ice. Soluble fractions were collected by centrifugation at 13,000 rpm for 10 min and analyzed by sodium dodecyl sulfate-polyacrylamide gel electrophoresis (SDS-PAGE). Human recoverin was electrophoretically extracted from gels. Fifty micrograms of extracted human recoverin was homogenized, emulsified with an equal volume of

FIG. 1. Immunofluorescence labeling for recoverin of gastric cancer tissue. After deparafinization of paraffin-embedded tumor sections obtained from gastric patients, the section was probed by antihuman recoverin monoclonal antibody (1 : 100 dilutions) and fluorescein isothiocyanate-labeled antimouse IgG (1 : 100 dilutions) as primary and secondary antibodies, respectively. Specific immunofluorescence labeling was photographed with a microscope (Olympus, Tokyo, Japan) using a blue filter set. *Bar* 10 μm

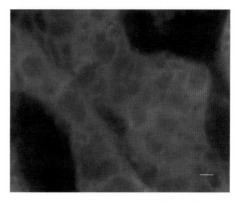

complete Freund's adjuvant (DIFCO), and immunized intraperitoneally to 4-week-old BALB/c mice (SLC, Shizuoka, Japan). Two weeks later and 3 days before fusion, human recoverin was injected with an equal volume of incomplete Freund's adjuvant (DIFCO). The immune response to human recoverin was determined by enzyme-linked immunoabsorbent assay.

Immunocytochemistry

Tumors, which had been surgically removed, were embedded in paraffin and sectioned at 3-μm thickness, mounted on subbed slides, and dried. The sections were incubated with antihuman recoverin mouse IgG at 1 : 100 in PBS for 1 h and rinsed with PBS three times for 5 min each. Then the sections were incubated with goat antimouse IgG labeled with fluorescein isothiocyanate at 1 : 500 in PBS containing 0.035% Tween 20 (T-PBS) at room temperature for 1 h. The sections were then rinsed three times with PBS for 5 min, and mounted on a slide glass using Vectashield fluorescence mounting medium (Vector Laboratories, Burlingame, CA, USA).

Results and Discussion

Recoverin exclusively expressed within photoreceptor cells and retinal bipolar cells is known to be a highly pathogenic molecule based upon the fact that immunization with purified recoverin induced high serum antibody titers toward recoverin, the activation of immunocompetent T cells, and photoreceptor degeneration in rats [12]. In CAR, aberrant expression of retina-specific recoverin in tumor cells is suggested to be a possible mechanism of the autoantibody production [8]. Similarly, it was found that a Purkinje cell antigen (Yo antigen) was recognized within the tumors in paraneoplastic cerebellar degeneration of patients with gynecological tumors [13]. Nevertheless, such an antigen was not detected in similar tumors obtained from individuals without neurological symptoms. Therefore, aberrant expressions of neuron-specific molecules in tumor cells seem to be a key causal mechanism in the degeneration of the target neuronal regions. However, in our recent study, we found that recoverin was aberrantly expressed in cell lines from various cancerous patients at a high incidence (21 out of 31 cell lines from various cancers) [9]. If our observation is correct, an additional mechanism causing an autoimmune reaction toward recoverin must be required for developing CAR retinal degeneration in addition to the aberrant expression of recoverin in tumor cells. To investigate whether such recoverin expression is present within tumor tissues obtained from cancerous patients rather than from established cancer cell lines, we studied the recoverin expression immunocytochemically in tumor tissues from three surgically treated patients with several clinical stages of gastric cancer. As shown in Fig. 1, immunofluorescence labeling of antihuman recoverin serum was recognized within the cytosol of gastric cancer cells, and its staining pattern was identical to that of the tumor cell lines described previously.

References

1. Lennon VA (1994) Paraneoplastic autoantibodies: the case for a descriptive generic nomenclature. Neurology 44:2236–2240
2. Sawyer RA, Selhorst JB, Zimmerman LE, et al. (1976) Blindness caused by photoreceptor degeneration as a remote effect of cancer. Am J Ophthalmol 81:606–613
3. Jacobson DM, Thirkill CE, Tipping SJ (1990) A clinical triad to diagnose paraneoplastic retinopathy. Ann Neurol 28:162–167
4. Polans AS, Buczylko J, Crabb J, et al. (1991) A photoreceptor calcium binding protein is recognized by autoantibodies obtained from patients with cancer-associated retinopathy. J Cell Biol 112:981–989
5. Kawamura S (1990) Rhodopsin phosphorylation as a mechanism of cyclic GMP phosphodiesterase regulation by S-modulin. Nature 362:855–857
6. Ohguro H, Ogawa K, Maeda T, et al. (1999) Cancer-associated retinopathy induced by both anti-recoverin and anti-hsc70 antibodies in vivo. Invest Ophthalmol Vis Sci 40:3160–3167
7. Maeda T, Maeda A, Maruyama I, et al. (2001) Mechanisms of photoreceptor cell death in cancer-associated retinopathy. Invest Ophthalmol Vis Sci 42:705–712
8. Polans AS, Witkowska D, Haley TL, et al. (1995) Recoverin, a photoreceptor-specific calcium-binding protein, is expressed by the tumor of a patient with cancer-associated retinopathy. Proc Natl Acad Sci USA 92:9176–9180
9. Maeda A, Ohguro H, Maeda T, et al. (2000) Aberrant expression of photoreceptor-specific calcium-binding protein (recoverin) in cancer cell lines. Cancer Res 60: 1914–1920
10. Maeda A, Ohguro H, Nabeta Y, et al. (2001) Identification of human antitumor cytotoxic T lymphocytes epitopes of recoverin, a cancer-associated retinopathy antigen, possibly related with a better prognosis in a paraneoplastic syndrome. Eur J Immunol 31:563–572
11. Sobin LH, Wittekind C (1997) International union against cancer: TNM classification of malignant tumors, 5th ed. Wiley-Liss, New York.
12. Adamus G, Ortega H, Witkowska D, et al. (1994) Recoverin: a potent uveitogen for the induction of photoreceptor degeneration in Lewis rats. Exp Eye Res 59:447–455
13. Furneaux HM, Rosenblum MK, Dalmau J, et al. (1990) Selective expression of Purkinje-cell antigens in tumor tissue from patients with paraneoplastic cerebellar degeneration. N Engl J Med 322:1844–1851

Cancer-Associated Retinopathy (CAR) is Effectively Treated by Ca^{2+} Antagonist Administration

Hiroshi Ohguro, Ikuyo Maruyama, Yoshiko Takano, Hitoshi Yamazaki, Tomomi Metoki, Futoshi Ishikawa, Yasuhiro Miyagawa, Kazuhisa Mamiya, and Mitsuru Nakazawa

Summary. Cancer-associated retinopathy (CAR) is a paraneoplastic ocular manifestation in which recoverin acts as an autoantigen recognized by sera from patients. Recently, we have found that CAR-like retinal dysfunction was produced by intravitreous administration of antirecoverin antibody in Lewis rat eyes. In the present study, nilvadipine, a Ca^{2+} antagonist recently found effective as a drug for inherited retinal degeneration, was evaluated for its effectiveness on the retinal degenerations in CAR using these models. Under medication with nilvadipine, the functional and morphological properties of the retinas were evaluated functionally and morphologically following antirecoverin antibody intravitreous injection into Lewis rats' eyes (six rats, 12 eyes in each experimental condition were used). Administration of nilvadipine to the antirecoverin antibody-treated rats caused significant improvement of the deterioration of ERG and normalization of rhodopsin phosphorylation. The present data indicated that antirecoverin antibody-induced retinal dysfunction can be effectively treated by systemic administration of nilvadipine.

Key words. Autoimmunity, Cancer-associated retinopathy, Recoverin, Rhodopsin phosphorylation, Electroretinogram

Introduction

It has been reported that the central nervous system could be the target of remote effects of malignant tumors that result in a variety of paraneoplastic syndromes, such as the Lambert-Eaton myasthenic syndrome, paraneoplastic cerebellar degeneration, and paraneoplastic sensory neuronopathy [1]. In terms of the pathogenesis of the paraneoplastic syndrome, it was suggested that an immune reaction toward antigens shared by the tumor cells and the nervous system is involved. In the ophthalmology field, paraneoplastic retinopathy is known to be associated with patients with malignancies and is called cancer-associated retinopathy (CAR). It is characterized by

---

Department of Ophthalmology, Hirosaki University School of Medicine, 5 Zaifucho, Hirosaki 036-8562 Japan

sudden and progressive visual loss, ring scotoma, photopsia, impairment of dark adaptation, and abnormalities of the a- and b-waves of the electroretinogram (ERG), like retinitis pigmentosa (RP). Among the underlying primary cancers, small-cell lung carcinoma has been reported most frequently. In most cases, CAR is diagnosed before an underlying primary cancer is discovered and such patients have a relatively better prognosis [2, 3].

On the basis of histopathological and immunological studies in patients with CAR, it was suggested that photoreceptor loss is primarily caused by an autoimmune reaction against a photoreceptor-specific, 23-kDa calcium-binding protein called recoverin [4]. Functionally, recoverin was identified as playing a major role in light and dark adaptation by regulating rhodopsin phosphorylation and dephosphorylation in a calcium-dependent manner [5, 6]. Aberrant expression of recoverin has been identified in the cancer cells from several cancerous patients including those with CAR [7], suggesting that aberrant expression of recoverin in cancer cells may trigger an autoimmune reaction. In addition, other retinal antigens including 65-kDa protein [3], 48-kDa protein [8], enolase (46-kDa protein) [9], and neurofilament (58-62 kDa, 145-kDa, and 205-kDa proteins) [10] are also recognized by some CAR patients' sera. Among these retinal autoantigens, we found that the 65-kDa protein was heat shock cognate protein 70 (hsc 70) [8], and that CAR-like retinal dysfunction was produced by intravitreal injection of antirecoverin antibody and antihsc 70 antibody in Lewis rats [10]. Therefore, we suggested that autoimmune reactions toward recoverin and hsc 70 proteins play significant roles in the molecular pathology of CAR.

Regarding the molecular mechanisms causing the retinal dysfunction by antirecoverin antibody, Adamus et al. reported that antirecoverin antibody was internalized into photoreceptor cells and induced apoptotic cell death in a retinal cell culture system [11] and that intravitreous administration of antibody against recoverin in Lewis rat eyes also caused apoptotic death of photoreceptor cells in vivo [12]. In addition, our group independently found that intravitreously administered antirecoverin antibody penetrated into photoreceptors, bound recoverin, and blocked the recoverin function that inhibits rhodopsin phosphorylation in a Ca^{2+}-dependent manner [13]. Therefore, on the basis of these observations, we speculate that the lack of the recoverin function by antirecoverin antibody causes enhancement of rhodopsin phosphorylation, resulting in the continuous opening of cyclic guanosine monophosphate (cGMP)-gated channels and the accumulation of intracellular Ca^{2+} within photoreceptor cells. If this is really involved in the molecular mechanisms in CAR, a decrease of the light-dependent rhodopsin phosphorylation levels by dark, or suppression of the increase of intracellular levels of Ca^{2+} by a Ca^{2+} antagonist may have a beneficial effect on the retinal dysfunction.

In the present study, in order to elucidate the effect of light on retinopathy in CAR, antirecoverin antibody-induced retinal dysfunction was evaluated under different illumination conditions and compared with phototoxic photoreceptor degeneration. Furthermore, we also examined the effects of Ca^{2+} antagonist and other drugs, corticosteroid and cyclosporin A, which were clinically used for human CAR patients, on the antirecoverin-induced retinal dysfunction in Lewis rats.

Materials and Methods

All experimental procedures were designed to conform to both the ARVO statement for Use of Animals in Ophthalmic and Vision Research and our own institution's guidelines. Unless otherwise stated, all procedures were performed at 4°C or on ice using ice-cold solutions.

Vitreous Injection of Antibodies in Lewis Rat

Intravitreous injection of antibody was performed as described by Ohguro et al. [14]. Briefly, a total of 5 μl of phosphate buffered saline (PBS) or antirecoverin IgG (5 μg) was administered into the vitreous cavity of a Lewis rat eye under anesthesia. The injection was performed with a 26-gauge Hamilton micro needle syringe through the sclera at a point 1 mm from the limbus to avoid puncture through the lens.

Drug Administration

Nilvadipine (5-isopropyl 3-methyl 2-cyano-6-methyl-4-(3-nytrophenol)-1,4-dihydro-3,5- pyridine dicarboxylate; Fujisawa Pharmaceutical, Tokyo, Japan) was dissolved in a mixture of ethanol:polyethylene glycol 400:distilled water (2:1:7) at a concentration of 0.1 mg/ml, diluted twice with physiological saline before use, and injected intraperitoneally (0.5 ml/kg) into anesthetized rats once a day for 3 weeks. In control rats, the same solution without nilvadipine (vehicle solution) was administered. Prednisolone (0.6 mg/kg/day) and cyclosporin A (10 mg/kg/day) dissolved in PBS and pure olive oil, respectively, were each administered by intramuscular injection.

Electroretinography

The anesthetized animals were kept in dark adaptation for at least 2 h in an electrically shielded room. The pupils were dilated with drops of 0.5% tropicamide. The scotopic ERG response was recorded as described by Ohguro et al. [14].

Statistical Analysis

The experimental data of ERG amplitudes ($n = 6$ rats, 12 eyes in each condition) are shown as mean ± SD. Significant differences between groups were found using the Mann-Whitney test with a significance level of $P < 0.05$ (*), $P < 0.01$ (**), or $P < 0.001$ (***).

Results and Discussion

In terms of therapy for patients with CAR and other paraneoplastic syndromes such as paraneoplastic cerebellar degeneration and Lambert-Eaton myasthenic syndrome, steroid administration, immunomodulation, and plasmapheresis have been clinically performed in conjunction with antineoplastic therapy [15–17]. For CAR, no definitive therapy has been established, although it has been reported that these treatments may be effective in some patients [3, 8]. Recently, it has been reported that lowering

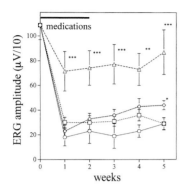

FIG. 1. Effects of corticosteroid or cyclosporin A administration on a scotopic electroretinogram (*ERG*) in Lewis rats intravitreously penetrated with antirecoverin IgG and antihsc 70 serum. A mixture of antihsc 70 (5 μg of IgG) and antirecoverin IgG (5 μg) was injected intravitreously into Lewis rat eyes, and these animals were intramuscularly administered either PBS (*circles*), prednisolone (0.6 mg/kg, *squares*), cyclosporin A (10 mg/kg, *diamonds*), or nilvadipine (0.025 mg/kg, *triangles*) every day during the next 2 weeks (marked by a *thick line* designated *medications*). During the 5-week period after the antibody penetration, scotopic ERG was recorded once a week. The b-wave amplitudes of the ERG were plotted

of intracellular Ca^{2+} by a Ca^{2+} antagonist and other drugs effectively suppresses the retinal neuronal apoptosis induced in some experimental animal models by ischemia-reperfusion [18] and intravitreal injection of N-methyl-D-aspartate (NMDA) [19]. These observations allowed us to speculate that similar effects by Ca^{2+} antagonist may be expected in our CAR model rats. In the present study, we found no significant effects of steroid or cyclosporin A on the antirecoverin antibody-induced retinal dysfunction. However, nilvadipine demonstrated remarkable effects on the antirecoverin antibody-induced retinal dysfunction (Fig. 1). At present, the precise mechanisms of nilvadipine have not been elucidated. However, we speculate that the lowering of intracellular Ca^{2+} levels by nilvadipine may inhibit the Ca^{2+}-dependent apoptotic pathway. Interestingly, Frasson et al. [20] recently reported rod photoreceptor rescue by lowering intracellular Ca^{2+} levels in photoreceptor cells using D-*cis*-diltiazem, a Ca^{2+} channel blocker in a different animal model of RP, rd mouse, in which the gene encoding cGMP phosphodiesterase is affected. However, recently, Bush et al. [21] reported that a Ca^{2+} antagonist, diltiazem, had no effects on the photoreceptor degeneration in the rhodopsin P23H rat. Therefore, it seems likely that Ca^{2+} channel blockers have protective effects on the retinal degeneration in some disease models but that these effects may be variable among different models, different species, disease stage, and different Ca^{2+} antagonists.

References

1. Hinton RC (1996) Paraneoplastic neurologic syndromes. Hematol Oncol Clin North Am 10:909–925
2. Sawyer RA, Selhorse JB, Zimmerman LE, et al. (1976) Blindness caused by photoreceptor degeneration as a remote effect of cancer. Am J Ophthalmol 81:606–613
3. Jacobson DM, Thirkill CE, Tipping SJ (1990) A clinical triad to diagnose paraneoplastic retinopathy. Ann Neurol 28:162–167
4. Polans AS, Buczylko J, Crabb J, et al. (1991) A photoreceptor calcium binding protein is recognized by autoantibodies obtained from patients with cancer-associated retinopathy. J Cell Biol 112:981–989
5. Kawamura S (1990) Rhodopsin phosphorylation as a mechanism of cyclic GMP phosphodiesterase regulation by S-modulin. Nature 362:855–857

6. Ohguro H, Rudonicka-Nawrot M, Palczewski K, et al. (1996) Structural and enzymatic aspects of rhodopsin phosphorylation. J Biol Chem 271:5215–5224
7. Polans AS, Witkowska D, Haley TL, et al. (1995) Recoverin, a photoreceptor-specific calcium-binding protein, is expressed by the tumor of a patient with cancer-associated retinopathy. Proc Natl Acad Sci USA 92: 9176–9180
8. Ohguro H, Ogawa K, Nakagawa T (1999) Recoverin and Hsc 70 are found as autoantigens in patients with cancer-associated retinopathy. Invest Ophthalmol Vis Sci 40: 82–89
9. Adamus G, Aptsiauri N, Guy J, et al. (1996) The occurrence of serum autoantibodies against enolase in cancer-associated retinopathy. Clin Immunol Immunopathol 78:120–129
10. Korngruth SE, Kalinke T, Dahl D, et al. (1986) Anti-neurofilament antibodies in the sera of patients with small cell carcinoma of the lung and with visual paraneoplastic syndrome. Cancer Res 46:2588–2595
11. Adamus G, Machnicki M, Seigel GM (1997) Apoptotic retinal cell death induced by antirecoverin autoantibodies of cancer-associated retinopathy. Invest Ophthalmol Vis Sci 38:283–291
12. Adamus G, Machnicki M, Fox DA, et al. (1998) Antibodies to recoverin induce apoptosis of photoreceptor and bipolar cells in vivo. J Autoimmun 11:523–533
13. Maeda T, Maeda A, Ohguro H, et al. (2001) Mechanisms of photoreceptor cell death in cancer-associated retinopathy. Invest Ophthalmol Vis Sci 42: 705–712
14. Ohguro H, Ogawa K, Maeda T, et al. (1999) Cancer-associated retinopathy induced by both anti-recoverin and anti-hsc70 antibodies in vivo. Invest Ophthalmol Vis Sci 40:3160–3167
15. Chalk CH, Murray NM, Spiro SG, et al. (1990) Response of the Lambert-Eaton myasthenic syndrome to treatment of associated small-cell lung carcinoma. Neurology 40:1552–1556
16. Anderson NE (1989) Anti-neuronal autoantibodies and neurological paraneoplastic syndromes. Aust N Z J Med 19:379–387
17. Hammack JE, Kimmel DW, Lennon VA, et al. (1990) Paraneoplastic cerebellar degeneration: a clinical comparison of patients with and without Purkinje cell cytoplasmic antibodies. Mayo Clin Proc 65:1423–1431
18. Takahashi K, Lam TT, Tso MO, et al. (1992) Protective effects of flunarizine on ischemic injury in the rat retina. Arch Ophthalmol 110:862–870
19. Osborne NN, DeSantis L, Chidlow G, et al. (1999) Topically applied betaxolol attenuates NMDA-induced toxicity to ganglion cells and the effects of ischaemia to the retina. Exp Eye Res 69:331–342
20. Frasson M, Sahel JA, Picaud S, et al. (1999) Retinitis pigmentosa: rod photoreceptor rescue by a calcium-channel blocker in the rd mouse. Nat Med 5:1183–1187
21. Bush RA, Kononen L, Sieving PA, et al. (2000) The effect of calcium channel blocker diltiazem on photoreceptor degeneration in the rhodopsin Pro213His rat. Invest Ophthalmol Vis Sci 41:2697–2701

Effects of Calcium Channel Blockers on Retinal Morphology and Function of rd Mouse

Yoshiko Takano, Hiroshi Ohguro, Hitoshi Yamazaki,
Ikuyo Maruyama, Tomomi Metoki, Futoshi Ishikawa,
Yasuhiro Miyagawa, Kazuhisa Mamiya, and Mitsuru Nakazawa

Key words. Retinal dystrophy, Retinitis pigmentosa, Electroretinogram, rd mouse

Introduction

Retinitis pigmentosa (RP) is a disease of inherited retinal degeneration characterized by nyctalopia, ring scotoma, and bone-spicule pigmentation of the retina. So far, no effective therapy has been available for RP. Several animal models with inherited retinal degeneration have been studied in order to elucidate the molecular pathology of RP, and to design an effective therapy for it. Frasson et al. [1] recently reported rod photoreceptor rescue by D-*cis*-diltiazem, a Ca^{2+} channel blocker in a different animal model of RP, rd mouse, in which the gene encoding cyclic guanosine monophosphate phosphodiesterase is affected. In addition, our group found that systemic administration of another type of Ca^{2+} channel blocker, nilvadipine, caused significant preservation of retinal morphology and function in Royal College of Surgeons (RCS) retinal degeneration [2]. These data suggest that the regulation of intracellular Ca^{2+} levels may be a possible therapy to prevent progressive retinal degeneration in RP and its animal models. So far, many Ca^{2+} channel blockers have been used in our clinical practice. Therefore, it seems very important to know which Ca^{2+} channel blocker is the most effective for retinal degeneration in RP.

In the present study, to answer this question, several kinds of Ca^{2+} antagonists used in clinical practice [3–5] were administrated to rd mice and then the retinal function by electroretinography (ERG) and morphology were investigated.

Materials and Methods

All experimental procedures were designed to conform to both the ARVO statement for Use of Animals in Ophthalmic and Vision Research and our own institution's guidelines.

Department of Ophthalmology, Hirosaki University School of Medicine, 5 Zaifucho, Hirosaki 036-8562, Japan

Anesthesia of the Animals

In the present study, 20-day-old rd mice reared in cyclic light conditions (12 h on/ 12 h off) were used. Mice were injected intramuscularly with a mixture of ketamine (80–125 mg/kg) and xylazine (9–12 mg/kg).

Drug Administration

Nilvadipine and nifedipine were dissolved in a mixture of ethanol: polyethylene glycol 400: distilled water (2:1:7) at a concentration of 0.1 mg/ml, diluted twice with physiological saline before use, and injected intraperitoneally (1.0 ml/kg) into mice every day early in the morning for 2 weeks. In the control, the same solution without nilvadipine or nifedipine (vehicle solution) was administered as described. Nicardipine and diltiazem were dissolved in phosphate-buffered saline (PBS) at 0.25 mg/ml and 1 mg/ml, respectively, and injected intraperitoneally (1.0 ml/kg) as described earlier. As a control, the same volume of a mixture of ethanol: polyethylene glycol 400: distilled water (2:1:7) or PBS was administered. Before administration, the pH of all drug solutions was adjusted to around 7.4.

Light Microscopy

Five rd mice were studied at the age of 20 days, weighing approximately 50 g for each control and four different Ca^{2+} antagonist administration conditions. Under deep anesthesia, animals were perfused through the initial portion of the aorta with 300 ml of 4% paraformaldehyde in 0.1 M PBS (pH 7.4), and the retinas were dissected out and embedded in paraffin. Posterior segments of eyes containing the optic disc at the center were cut from the enucleated eyes and embedded in paraffin. Retinal sections were cut vertically through the optic disc at 4-μm thickness from nasal to temporal, mounted on subbed slides, and dried. The sections were processed with hematoxylin-eosin staining after deparaffinization with graded ethanol and xylene solutions.

Electroretinography

Five mice were studied at the age of 20–30 days, weighing approximately 50 g for each control and four different Ca^{2+} antagonist administration conditions. Under anesthesia, the rats were dark adapted for overnight. The pupils were dilated with drops of 0.5% tropicamide. ERGs were recorded with a contact electrode equipped with a suction apparatus to fit on the cornea (Kyoto Contact Lens, Kyoto, Japan). A grounding electrode was placed on the ear. A response evoked by a white flash (3.5×10^2 lux, 200-ms duration) was recorded by a Neuropack (MES-3102, Nihon Kohden, Tokyo, Japan).

Results and Discussion

Ca^{2+} antagonists, which have been widely used as treatments for systemic hypertension, inhibit the entry of calcium ion intracellularly, relax vascular smooth muscle cells, and increase regional blood flow in several organs [3, 4]. As major Ca^{2+} antago-

A

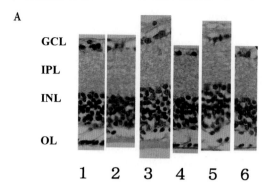

GCL;ganglion cell layer,IPL;inner plexiform layer,
INL;inner nuclear layer,OL;outer layer(ONL+OPL+OS)

1; vehicle(NaCl) 2; nicardipine 3; diltiazem
4;vehicle(ethanol)   5; nilvadipine   6; nifedipine

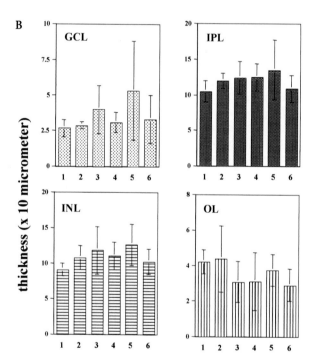

thickness (x 10 micrometer)

GCL; ganglion cell layer, IPL; inner plexiform layer
INL; inner nuclear layer, OL; outer layer

1; saline, 2; nicardipine, 3; diltiazem
4; EtOH/water/polyethylene glycol, 5; nilvadipine
6; nifedipine

nists, a dihydropyridine (DHP) derivative such as nifedipine, nicardipine, and nilvadipine; a benzothiazepine derivative such as diltiazem; and a phenylalkylamine derivative such as verapamil are known to be used in our clinical practices [5]. On the basis of kinetics and voltage-dependent properties, the Ca^{2+} channels have been classified into two groups; one is activated by small depolarizations and is then subsequently inactivated, and is called the low-voltage-activated (LVA) Ca^{2+} channel, whereas the other is activated by larger depolarizations and shows little inactivation, and is called the high-voltage-activated (HVA) Ca^{2+} channel. On the basis of pharmacological properties, HVA Ca^{2+} channels can be separated further into four types (L, N, P/Q, and R) [6, 7].

In terms of retinal Ca^{2+} channels, it was revealed that currents of L-type HVA Ca^{2+} channels in the vertebrate photoreceptors were recognized and were sensitive to DHPs [8–10]. Recently, a novel Ca^{2+} channel gene, *CACN1F*, encoding α_{1F}, a retina-specific α_1 subunit of L-type HVA Ca^{2+} channels, was identified; mutations in *CACN1F* cause incomplete X-linked congenital stationary night blindness (CSNB2) [11, 12]. On the basis of these observations, we can reasonably speculate that these drugs may react with retinal L-type HVA Ca^{2+} channels, and may presumably prevent photoreceptor cell death. In fact, it was recently reported that the rod photoreceptor of rd mouse was rescued by D-*cis*-diltiazem [1]. However, in contrast, Bush et al. [13] reported that the Ca^{2+} antagonist D-*cis*-diltiazem had no effects on the photoreceptor degeneration in

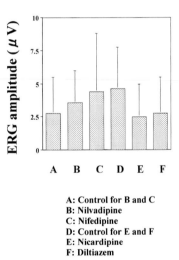

Fig. 2A–F. Effects of Ca^{2+} antagonists on scotopic electroretinography (*ERG*) in rd mouse. rd mouse treated with D-*cis*-diltiazem (*F*), nifedipine (*C*), nicardipine (*E*), nilvadipine (*B*), or their vehicle solution (*A*, *D*), respectively, every day from 3 weeks old. During the 2 weeks after the operation, scotopic ERG was recorded once a week. ERG measurements were performed in ten eyes (five mice) in each condition, and the a-wave and b-wave amplitudes were plotted

A: Control for B and C
B: Nilvadipine
C: Nifedipine
D: Control for E and F
E: Nicardipine
F: Diltiazem

Fig. 1A, B. Effects of several Ca^{2+} antagonists on the thickness of the retinal layers of rd mouse. Hematoxylin-eosin staining of retinal sections at 1 mm from the optic disc from rd mouse eyes at the age of 4 and 5 weeks treated with D-*cis*-diltiazem, nifedipine, nicardipine, nilvadipine, or their vehicle solutions. Photographs of the sections were taken (A), and each of the retinal layers was measured at temporal and nasal points 1 mm from the optic disc from five different retinas (two different points per one eye, total 10 different points) and plotted (B). *GCL*, ganglion cell layer; *IPL*, inner plexiform layer; *INL*, inner nuclear layer; *OL*, outer layer; *OPL*, outer plexiform layer; *ONL*, outer nuclear layer; *OS*, outer segment. *$P < 0.01$ (Mann-Whitney test)

the rhodopsin P23H rat. More recently we found that nilvadipine preserved photoreceptor cell function and structure in RCS rat, but its retinal degeneration was still progressive. Therefore, it seems likely that Ca^{2+} channel blockers have protective effects on the retinal degeneration in some disease models, but these effects may be variable among different models, different species, disease stage, and different Ca^{2+} antagonists.

In the present study, to elucidate which Ca^{2+} antagonist is the most beneficial for preservation in the retinal degeneration, we studied the drug effects of several Ca^{2+} antagonists on the retinal degeneration of rd mouse to evaluate their efficacy. Four Ca^{2+} antagonists, diltiazem, nicardipine, nilvadipine, or nifedipine, were intraperitoneally administered to rd mice (20 days old) for 1 week and thereafter retinal morphology and functions were analyzed. We found that systemic administration of these Ca^{2+} antagonists caused no preservation of retinal morphology (Fig. 1) and functions of ERG responses (Fig. 2) in rd mouse during a relatively advanced period of the retinal degeneration. On the basis of these data, it is suggested that Ca^{2+} antagonists are not beneficial for the preservation of photoreceptor cells in the relatively advanced stage of retinal degeneration in rd mouse. Therefore, we must perform a similar study using an earlier stage animal model of retinal degeneration as next our project.

References

1. Frasson M, Sahel J, Fabre M, et al. (1999) Retinitis pigmentosa: rod photoreceptor rescue by a calcium-channel blocker in the rd mouse. Nat Med 5:1183–1187
2. Yamazaki H, Ohguro H, Maeda T, et al. (2002) Preservation of retinal morphology and functions in royal college surgeons rat by nilvadipine, a Ca^{2+} antagonist. Invest Ophthalmol Vis Sci 43:919–926
3. Hof RP (1983) Calcium antagonist and the peripheral circulation: differences and similarities between PY108–068, nicardipine, verapamil and D-cis-diltiazem. Br J Pharmacol 78:375–394
4. Ohtsuka M, Yokota M, Kodama I, et al. (1989) New generation dihydropyridine calcium entry blockers: in search of greater selectivity for one tissue subtype. Gen Pharmacol 20:539–556
5. Triggle DH (1981) Calcium antagonist: basic chemical and pharmacological aspects. In: Weiss GB (ed) New perspectives on calcium antagonists. American Physiological Society, Bethesda, pp 1–21
6. Tsein RW, Ellinor PT, Horne WA (1991) Molecular diversity of voltage-dependent Ca^{2+} channels. Trends Pharmacol Sci 12:349–354
7. Miller RJ (1992) Voltage sensitive Ca^{2+} channels. J Biol Chem 267:1403–1406
8. Ohtsuka M, Ono T, Hiroi J, et al. (1983) Comparison of the cardiovascular effect of FR34235, a new dihydropyridine, with other calcium antagonists. J Cardiovasc Pharmacol 5:1074–1082
9. Tomita K, Araie M, Tamai Y, et al. (1999) Effects of nilvadipine, a calcium antagonist, on rabbit ocular circulation and optic nerve head circulation in NTG subjects. Invest Ophthalmol Vis Sci 40:1144–1151
10. Chik CL, Li B, Negishi T, et al. (1999) Ceramide inhibits L-type calcium channel currents in rat pinealocytes. Endocrinology 140:5682–5690
11. Bech-Hansen NT, Naylor MJ, Maybaum TA, et al. (1998) Loss-of-function mutations in a calcium-channel alphal-subunit gene in Xp 11.23 cause incomplete X-linked congenital stationary night blindness. Nat Genet 19:264–267

12. Strom TM, Nyakatura G, Apfelstedt-Sylla E, et al. (1998) An L-type calcium-channel gene mutated in incomplete X-linked congenital stationary night blindness. Nat Genet 19:260–263
13. Bush RA, Kononen L, Machida S, et al. (2000) The effect of calcium channel blocker diltiazem on photoreceptor degeneration in the rhodopsin Pro213His rat. Invest Ophthalmol Vis Sci 41:2697–2701

Chemically Induced Mutations in the Mouse that Affect the Fundus and Electroretinogram

LAWRENCE H. PINTO[1], NICHOLAS GRABOWSKI[1], WARREN A. KIBBE[1],
ANDREW LOTERY[2], STEPHEN LUMAYAG[1], ROBERT MULLENS[2],
EDWIN STONE[2], SANDRA SIEPKA[1], MARTHA H. VITATEMA[1],
VAL SHEFFIELD[3], and JOSEPH S. TAKAHASHI[1]

Key words. Forward genetics, Mutagenesis, Gene discovery

C57BL/6J ("Black 6") mice were mutagenized with a chemical mutagen (ENU) that induces point mutations and were bred for three generations to render recessive mutations homozygous. Both first- (Gl, dominant screen) and third- (G3, recessive screen) generation mice were tested in two ways. First, the retinogram in response to nine different stimuli was measured automatically after dark adaptation and during presentation of a steady light to light adapt the retina. Second, the fundus was photographed and analyzed for abnormalities. The results of screening 177 G3 mice from 31 Gl founders are as follows. One G3 mouse was found to have normal a and b waves, but no c wave after retesting. Another G3 mouse produced no measurable electroretinogram (ERG) and both of its fundi appeared very abnormal; the heredity of these traits are now being tested. We examined the fundus of 610 G3 mice, 100 G2 mice, and 79 Gl mice. Twenty-four G3 mice had an abnormal fundus appearance but their ERGs were indistinguishable from normal. For 19 G3 mice, the branching pattern of the retinal vessels was abnormal in one eye only; for 2 mice from the same G1 pedigree, the branching pattern of the retinal vessels was abnormal in both eyes, and for 5 mice the fundus or the optic disk appeared to have abnormal white blotches that are now being characterized histologically. Thus, screening for visual phenotypes among a relatively small number of mutagenized mice has produced a promising number of "leads," demonstrating the feasibility for using this approach to generate mouse models of human diseases and to use "genetic dissection" to help understand the visual system. We are now in the process of increasing the rate of screening to 10,000 G3 mice per year and are developing a rapid test for visually evoked potentials of the cortex. Results will be posted on the web site http://genomics.northwestem.edu/. Members of the community will be offered the opportunity to create a "profile" of the phenotype of a mutant that would be of interest to them, and if such a mutant occurs

[1] Center for Functional Genomics and Department of Neurobiology and Physiology, Northwestern University, Evanston, IL 6020, USA
Departments of [2] Ophthalmology and [3] Pediatrics, University of Iowa, Iowa City, IA 52242, USA

in the process of screening over the next 4 years, the requester will be informed immediately by e-mail and given the information on how to obtain that mutant. All mice
will be screened for nine other nonvision phenotypes so that the specificity of the
visual defects will be known.

Unitary Event Analysis of Synchronous Activities in Cat Lateral Geniculate Nucleus (LGN)

Akio Hirata[1,2], Pedro E. Maldonado[3], Charles M. Gray[4], and Hiroyuki Ito[1,2]

Key words. Synchronization, Oscillation, Multiunit recording

Introduction

Several studies have shown that cells in the visual cortex (A17) of the cat display stimulus-evoked oscillatory firings (30–60 Hz) and synchronization among multiple cells [1]. Ito et al. [2] reported that the cells in the lateral geniculate nucleus (LGN) also showed stimulus-evoked synchronous oscillatory firings of a higher frequency range (60–100 Hz). The difference in these frequency ranges suggested that the oscillatory activity in the cortex was not likely to derive from that in the LGN. In addition, Gray et al. [3] reported the transient nature of the synchronization in the cortex, that is, the transition between asynchronous and synchronous states, could occur within a trial duration. For further comparison of the synchronization between the two areas, we investigated whether such nonstationarity exists also in the synchronization in the LGN. Since the traditional method of cross-correlogram loses any nonstationarity within the trial duration by temporal averaging, we adopted a novel method, unitary event (UE) analysis, which was originally applied to the higher area (motor cortex) [4].

Materials and Methods

Electrophysiological recordings were performed on 9 adult cats using previously published techniques [1]. The subjects were maintained under general anesthesia using

[1] Faculty of Engineering, Kyoto Sangyo University, Motoyama Kamigamo, Kita-ku, Kyoto 603-8555, Japan
[2] Core Research for Evolutional Science and Technology (CREST), Japan Science and Technology Corporation (JST), Kawaguchi Center Building, 4-1-8 Hon-machi, Kawaguchi, Saitama 332-0012, Japan
[3] Instituto de Ciencias Biomédicas, Facultad de Medicina, Universidad de Chile, Santiago, Chile
[4] Center for Computational Biology, Montana State University, 1 Lewis Hall Bozeman, MT 59717, USA

halothane or isoflurane in a mixture of N_2O and O_2 (2:1) throughout surgery and during the recordings. Stationary light spots were presented (0.5–3 s) at the receptive fields on the monitor with a refresh rate of 80 Hz placed at a distance of 57 cm. Activities of the neuron were simultaneously recorded by two tetrodes (separated by 500 µm) in all layers of the LGN. Units were spike sorted from the multiunits by the cluster-cutting algorithm.

The previous analysis of the cross-correlogram showed that 84 pairs out of 847 pairs were judged as significantly synchronized (see [5] for statistical criteria). The UE analysis was applied to those 84 sets of pairs for further examination of non-stationarity (see [6] for the method of UE analysis). First, the peak of the cross-correlogram was detected in a precision of 1 ms. In the pair of spike trains in each trial, the spike pairs of the two units having this time delay were picked up throughout the trial duration (coincident events, CEs). The number of the raw CEs within a small moving window (101 ms) was averaged over all the trials. The number of accidental CEs in the null hypothesis of independent firings obeys the binomial distribution computed by the firing rates of the two units. The mean of the distribution gives the predicted number of CEs in the null hypothesis. The number of raw CEs was judged as significant (UE) when it exceeded the 95% confidence limit of the binomial distribution (rejection of the null hypothesis). We repeated such statistical tests by sliding the window on a 1-ms basis over the trial duration. The nonstationarity of the number of CEs was quantified by computing the modulation index (MI). First, we subtracted the number of predicted CEs from the number of raw CEs. Then, within a specific interval (stimulus duration etc.), we computed the maximum and the minimum of the modulation. Finally, the difference between the two values was divided by the average number of the predicted CEs to compute the MI^{test}. Similarly, we computed the MI^{shift} for all possible trial-shifted combinations. The test data were judged as having a significant temporal modulation when the MI^{test} was larger than the maximum of the MI^{shift}s.

Results

A typical example showing a significant temporal modulation of the number of CEs is shown in Fig. 1. The UEs appeared transiently during the limited interval of the on-response. The number of CEs started to decrease to the predicted value even when the rates of the two units remained high. In the off-response, the two units fired almost independently. Those correlation structures were completely eliminated by the trial shuffling. On the other hand, the other example showed a stationary synchronization; that is, the UEs appeared continuously during stimulus presentation (figure not shown). The different types of the synchronization were classified on the basis of the two criteria, the significance of the number of the CEs, and the significance of the MI (Table 1). About 40% of the total samples showed the UEs and 67% of the cases having the UEs showed a significant temporal modulation. Some unit pairs were excluded from the analysis because of not enough spike counts.

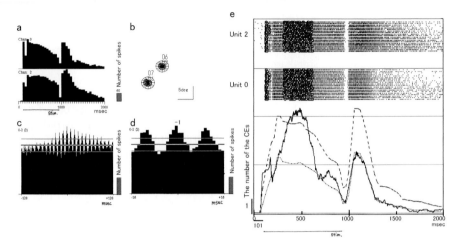

FIG. 1a–e. Typical example of a significant temporal modulation of the number of coincident events (*CEs*). **a** Poststimulus time histogram of the two single units recorded by the different tetrodes. The on/off-response and the response were observed. **b** Receptive field (RF) of each unit. The spots were simultaneously presented in two RFs. **c** The cross-correlogram between the two units during the on-responses showing a synchronization. Both units showed vigorous oscillatory firings (auto-correlogram not shown). **d** Magnified view of the cross-correlogram. The peak at −1 ms suggests that unit 2 tended to fire 1 ms prior to unit 0. **e** Raster plot of each unit and the unitary events (UEs). *Solid line*, the number of CEs; *dotted line*, predicted number; *dashed line*, 95% confidence limit computed by the binomial distribution of the null hypothesis. The *open circles* on the raster plot represent CEs judged to be significant (UEs). The *UEs near the onset* are an artifact of the on-response transient. *Horizontal bar*, the size of the moving window (101 ms); *vertical bar*, scale bar (unity) of the number of CEs

TABLE 1. The classification of the coincident events (CEs) of the unit pairs

Significance of CEs	*Nonstationarity in CE number*		Total
	Significant	Not significant	
Signifcant	16	8	24 (38%)
Not significant	8	32	40 (62%)
			64

All data are the number of the unit pairs

Discussion

A large portion of the unit pairs having the UEs showed a nonstationary modulation of the number of CEs. It is probable that the synchronous oscillation derives from the retinal inputs [7]. What is the origin of the nonstationarity in the synchronization? It is known that there is little lateral connection within the LGN. One possibility is that nonstationarity exists already in the input from the retina. The other possibility is that the feedback from the cortex modulates the synchronization. In addition to the afferent inputs from the retina, the LGN receives massive feedback inputs from the cortex. What is the functional role of the nonstationarity of the UEs? To exam this question further, we need to analyze the change of the properties of the UEs under different stimulus contexts (spot, two short bars, single long bar).

Conclusion

UE analysis suggested that the synchronization between the cells in the LGN showed a nonstationary modulation within the trial duration even under the presentation of the stationary spot stimulus. The modulation of the number of CEs showed an apparently distinct temporal profile from that of the firing rates.

References

1. Gray CM, Viana Di Prisco G (1997) Stimulus-dependent neuronal oscillations and local synchronization in striate cortex of the alert cat. J Neurosci 17:3239–3253
2. Ito H, Gray CM, Viana Di Prisco G (1994) Can oscillatory activity in the LGN account for the occurrence of synchronous oscillations in the visual cortex? Abstr Soc Neurosci 20:134
3. Gray CM, Engel AK, Koenig P, et al. (1992) Synchronization of oscillatory neuronal responses in cat striate cortex: temporal properties. Vis Neurosci 8:337–347
4. Riehle A, Gruen S, Diesmann M, et al. (1997) Spike synchronization and rate modulation differentially involved in motor cortical function. Science 278:1950–1953
5. Maldonado PE, Friedman-Hill S, Gray CM (2000) Dynamics of striate cortical activity in the alert macaque: fast time scale synchronization. Cereb Cortex 10:1117–1131
6. Gruen S (1996) Unitary joint-events in multiple-neuron spiking activity. Detection, significance, and interpretation. Verlag Harri Deutsch, Frankfurt, pp 45–46
7. Castelo-Branco M, Neuenschwander S, Singer W (1998) Synchronization of visual responses between the cortex, lateral geniculate nucleus, and retina in the anesthetized cat. J Neurosci 18:6395–6410

Functional Mapping of Neural Activity in the Embryonic Avian Visual System: Optical Recording with a Voltage-Sensitive Dye

Naohisa Miyakawa[1], Katsushige Sato[1], Hiraku Mochida[1,2], Shinichi Sasaki[1], and Yoko Momose-Sato[1]

Summary. We optically investigated the developmental patterns of signal propagation and synaptic transmission in the embryonic chick visual system. Optic tectum preparations with the optic nerve intact were dissected from E11 embryos. Optic nerve stimulation elicited fast spike-like and slow signals in the contralateral tectum. Application of Cd^{2+} to the extracellular solution or removing extracellular Ca^{2+} eliminated the slow signals. Application of tetrodotoxin to the bathing solution eliminated both the fast and slow signals. These results indicate that the fast and slow signals correspond to the action potential and excitatory postsynaptic potential, respectively. Furthermore, we constructed color-coded spatiotemporal maps of neural activity induced by optic nerve stimulation. We suggest that, in comparison with anatomical information, the optical response areas correspond to the visual sensory nuclei including the ectomammilary nucleus and the optic tectum, and to the nucleus of the centrifugal pathway, the isthmo optic nucleus.

Key words. Optical recording, Chick optic tectum, Synaptic transmission, Development, Voltage-sensitive dyes

Introduction

The optic tectum in birds is indispensable for visual information processing, especially motion detection. Unlike those in the mammalian brain, many important visual inputs take the retinotectal pathway, whereas fewer take the retinothalamic pathway. Although the optic tectum plays a critical role in the well-developed avian visual system, its functional development has not yet been clarified because of the small size, fragility, and topology of immature or newly matured neurons in the embryonic brain.

Optical recording techniques with voltage-sensitive dyes (VSD) are a powerful tool for studying neural activity in the embryonic nervous system [1]. We applied an optical technique to the embryonic chick visual system, and succeeded in recording neural activity from multiple regions in the optic tectum.

[1] Department of Physiology and Cell Biology, Tokyo Medical and Dental University, Graduate School and Faculty of Medicine, 1-5-45 Yushima, Bunkyo-ku, Tokyo 113-8519, Japan
[2] Department of Physiology, University of Oslo School of Medicine, Oslo, Norway

Materials and Methods

Whole brain preparations with the optic nerve intact were dissected from embryonic day 11 (E11, Hamburger-Hamilton stage 37 [2]) chick embryos in cold-oxygenated Ringer's solution (Fig. 1a). It has been reported that the excitatory postsynaptic potential (EPSP) is first detected with electrophysiological means at this stage [3]. The cerebellum and caudal half of the optic tectum were removed (Fig. 1b, left). The preparation was stained by incubating it for 20 min in a Ringer's solution containing 0.2 mg/ml of a voltage-sensitive merocyanine-rhodanine dye, NK2761. Excess (unbound) dye was washed away, and the preparation was recovered for an hour at room temperature before recording. The stained preparation was pinned to a silicone rubber bottom in a recording chamber with the rostral side up (Fig. 1b, right). The chamber was set on an upright microscope. Light from a 300-W tungsten–halogen lamp was passed through a heat-cut filter and a 701 ± 38-nm band-pass filter and focused on the preparation. A square current pulse was delivered to the optic nerve with a suction electrode. The image of the preparation was magnified x 6.67–10, and changes in light intensity transmitted from the preparation were recorded simultaneously with a 1020-element photodiode array mounted on an upright microscope. The outputs of each detector were passed to amplifiers and recorded simultaneously. Experiments were carried out at room temperature, 26–28°C.

Results

Optical Signals Elicited by Optic Nerve Stimulation

Figure 1c illustrates an example of multiple-site optical recording of neural activity induced by optic nerve stimulation in an E11 optic tectum preparation. The detected optical signals showed a wavelength dependence with an isosbestic point at 640 nm (data not shown), indicating that they are related to transmembrane voltage changes.

The neural activity-related signals detected from the dorsomedial tectum showed a monophasic spike-like pattern (site 3). On the other hand, the signals detected from the ventromedial (site 1) and lateral (sites 4–6) regions exhibited a biphasic waveform. The first component was a fast spike-like signal, while the second component was a delayed long-lasting slow signal. The slow signals were reduced, or eliminated, by lowering the external Ca^{2+} concentration (data not shown), and they also were eliminated by addition of Cd^{2+} (1 mM) to the bathing solution (Fig. 1d). Both the fast and slow signals were completely eliminated by application of 2 μM tetrodotoxin (TTX). These results indicate that the fast spike-like signals correspond to action potentials, and that the slow signals correspond to EPSPs. The results also indicate that the detected action potential is dependent on sodium channel activity.

Spatiotemporal Patterns of Neural Activity

Figure 2 illustrates color-coded maps of spatiotemporal patterns of neural activity induced by optic nerve stimulation in the E11 chick optic tectum. Optical signals first appeared in the ventromedial part of the optic tectum with a 20-ms delay after stimulation, and propagated to the entire region of the optic tectum within the

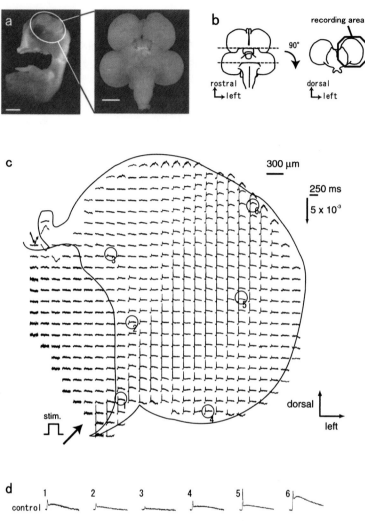

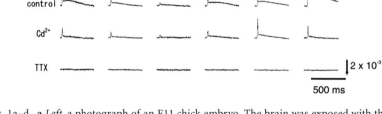

FIG. 1a–d. **a** *Left*, a photograph of an E11 chick embryo. The brain was exposed with the optic nerve intact (*bar* 5 mm). *Right*, a ventral view of a whole brain stained with a voltage-sensitive dye, NK2761 (*bar* 2 mm). **b** A schematic drawing of an optic tectum preparation. The brain was sliced along the *dotted line* (*left*), and pinned to a chamber with the rostral side up (*right*). **c** Multiple-site optical recording of neural activity induced by right optic nerve stimulation. The signals were detected from the left optic tectum. **d** Enlarged signals detected from six regions of the tectum indicated in **c**. The recordings were made in normal Ringer's solution (*top row*), a 1-mM Cd^{2+}-containing solution (*middle row*), and a 2-μM tetrodotoxin (*TTX*)-containing solution (*bottom row*)

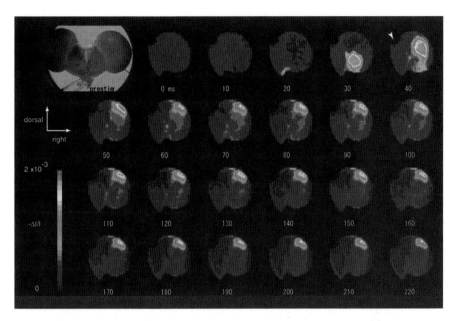

FIG. 2. Spatiotemporal color-coded maps of neural activity induced by optic nerve stimulation

succeeding 20 ms. The optical signals were most prominent in the dorsolateral region. A small, but significant, component also propagated along the medial portion of the tectum to the dorsomedial region (indicated by a white arrowhead). The signals persisted in the tectum for more than 200 ms, especially in the dorsolateral area.

Discussion

In the present study, we applied an optical technique with a voltage-sensitive dye to embryonic optic tectum preparations, and succeeded in recording neural activity simultaneously from multiple regions. As shown in Fig. 1c, when we stimulated the optic nerve, large, long-lasting optical signals were detected from the ventromedial (e.g., site 1) and dorsolateral (e.g., site 6) areas. In comparison with anatomical information [3], the area around site 1 seems to correspond to the ectomammilary nucleus, and the area around site 6 to the dorsolateral part of the optic tectum. From these areas, two types of optical signals were identified: one was a fast spike-like signal and the other was a delayed slow signal. From the results that the slow signal was blocked by Ca^{2+}-free or Cd^{2+}-containing Ringer's solution, and considering that the optic nerve bundles mainly contain sensory nerve fibers, it is reasonable to interpret the fast signals as corresponding to the orthodromic action potentials evoked in the sensory nerves and the slow signals as reflecting the postsynaptic potentials evoked in the postsynaptic neurons. Considering that the fast signal detected from site 6 was reduced slightly in amplitude by Ca^{2+}-free or Cd^{2+}-containing Ringer's solution, some fraction of the postsynaptic firing component might also contribute to the fast optical signals.

On the other hand, only spike-like signals were detected from the dorsomedial region (Fig. 1c, site 3). This spike-like signal could be either the firing of sensory (afferent) nerve fibers with no synaptic connection at this developmental stage (E11), or the antidromic firing of centrifugal (efferent) nerve fibers. We suggest that, compared with morphological data [3], the latter possibility is more plausible and that the area around site 3 corresponds to the ithmo optic nucleus.

References

1. Momose-Sato Y, Sato K, Kamino K (2001) Optical approaches to embryonic development of neural functions in the brainstem. Prog Neurobiol 63:151–197
2. Hamburger V, Hamilton HL (1951) A series of normal stages in the development of the chick embryo. J Morphol 88:49–92
3. Rager G (1976) Morphogenesis and physiogenesis of the retino-tectal connection in the chicken: II. The retino-tectal synapses. Proc R Soc Lond B Biol Sci 192:353–370

Interactive Vision: A New Columnar System in Layer 2

Noritaka Ichinohe and Kathleen S. Rockland

Key words. Columnar organization, Visual cortex, Retrosplenial cortex, Zinc-rich cortical terminal, Dendritic bundle

Introduction

In a review from 1994 by Churchland et al. [1], an interesting distinction was made between two approaches to the problem of how we see. One, equated with the more traditional approach, was called the theory of pure vision, according to which the visual system operates to create a detailed replica or representation of the visual world. This is accomplished independently of other sensory modalities and independently of previous learning, goals, planning, and motor execution. The alternative "interactive vision," by contrast, acknowledges and emphasizes the interdependence.

Since 1994, the shift toward "interactive vision" has continued, with major consequences for our ideas of visual cortical organization and function. Among the factors influencing the change in viewpoint are dynamic regulation of receptive field properties in V1, adaptation effects on neuronal responsivity, and evidence that levels of attention and arousal are important even in V1 (for one recent review, see [2]). From another quarter, several groups have now shown that the response latencies of neurons in different cortical visual areas are not in accord with a strict serial organization, and that complex perceptual processes that occur over hundreds of milliseconds are routinely accompanied by coactivity of a large number of visual areas including V1 (reviewed in [3]). All in all, then, the current view favors more the idea that "individual V1 neurons are not static filters," but instead clearly respond in a context-dependent and goal-oriented manner, with visual cortex functioning in a dynamic and cooperative mode as part of a distributed network.

At a connectional level, the neuroanatomical data strongly support the concept of a more interactive, dynamic V1. Multiple, parallel pathways interconnect V1 with the lateral geniculate nucleus, pulvinar, and (indirectly) superior colliculus, and with

Laboratory for Cortical Organization and Systematics, Brain Science Institute, RIKEN, 2-1 Hirosawa, Wako, Saitama 351-0198, Japan

several extrastriate cortical areas, including in the parahippocampal gyrus. A vivid indication that V1 is not a sequestered area is the report of inputs from auditory and parietal association areas to area V1 in primates, in the peripheral field representation [4, 5]. The amygdala is known to project to V1.

In the revised view of visual cortex, the role of thalamocortical connections is also undergoing reevaluation, with renewed appreciation of (1) the multiplicity of parallel circuits, and (2) the conjunctive importance of other connections, particularly cortical connections. As a consequence, it may be necessary to more fully consider other routes of intercortical processing, rather than the classical emphasis on interlaminar relays through layer 4.

The next part of this brief review, in fact, presents newer findings of a honeycomb-like specialization in layer 2 and adjacent layer 1. The honeycomb mosaic occurs widely in several species and areas, but has been best characterized in rat visual cortex [6]. It also occurs in the granular retrosplenial cortex (GRS), where it merges with a system of parvalbumin-immunoreactive (PV-ir) dendritic bundles in layer 1 [7]. One implication of these results is that, in neocortex, as in limbic cortices, layer 2 has a distinct and prominent role.

Rat Retrosplenial Cortex

The GRS in rats has a highly modular organization in layer 1, consisting of patches, 30–100 µm in width, of apical dendrites. These originate from callosally projecting pyramidal neurons in layer 2, as shown by retrograde labelling with Fluorogold, and can be demonstrated by intracellular filling with Lucifer yellow, Golgi preparations [8], and microtubule-associated protein 2 (MAP2) immunohistochemistry [7] (Fig. 1A). Tracer studies further demonstrate that thalamic terminations from the anteroventral (AV) nucleus specifically target the dendritic bundles, while intracortical projections terminate in the interbundle spaces [8].

Recently, we reported that dendrites of PV-ir, putative γ-aminobutyric acid (GABA)ergic interneurons, also form bundles in layer 1 (Fig. 1D), and that these comingle with the bundles of pyramidal cell apical dendrites. As additional evidence of modular specificity, there are high levels of both N-methyl-D-aspartate receptor 1 (NMDAR1; Fig. 1C) and GABAaα3 subunit (not shown), which are preferentially localized to the PV-ir dendritic bundles (Fig. 1D).

Calretinin (CR) immunoreactivity (reflecting CR-ir terminal-like puncta) also shows a periodic pattern in layer 1 (Fig. 1B). In contrast to NMDAR1 and GABAaα3, this is complementary to the MAP2-ir bundles (Fig. 1A,B), and perhaps targets another dendritic subsystem consisting of apical dendritic tufts from layer 3 and 5 pyramidal neurons. The origin of CR-ir terminal-like structures remains to be determined, but may be at least in part from CR-ir cell bodies known to exist in the laterodorsal thalamic nucleus, which projects to the GRS.

What is the functional significance of this modularity? The close association of PV-ir and apical dendrites together with thalamic terminations is suggestive of a specialized complex, possibly related to feedforward inhibition. Parallel microcircuitry in layer 1 is also suggested. That is, AV thalamocortical terminals would preferentially target tufts of layer 2 pyramidal neurons and corticocortical terminals, the tufts of layer 3 and 5 pyramidal neurons. The high levels of NMDAR1 would be consistent

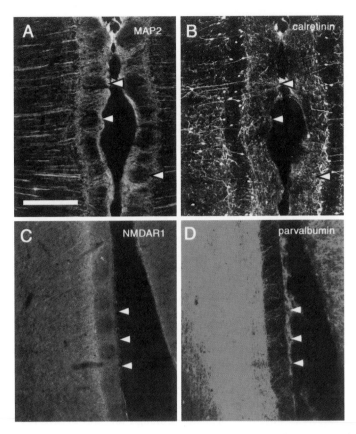

Fig. 1A–D. Coronal section through the granular retrosplenial cortex showing a patchy pattern in microtubule-associated protein 2 (*MAP2*), calretinin (CR), *N*-methyl-D-aspartate receptor 1 (*NMDAR1*), and parvalbumin (PV; confocal images of two different sections; color merge not shown). A and B Double immunofluorescence showing complementarity of MAP2 (**A**) and CR (**B**). CR immunoreactivity is lower in the MAP2-immunoreactive (ir) dendritic bundles and higher in the interbundle spaces, especially in layer 1a. C and D Double immunofluorescence for NMDAR1 (**C**) and PV (**D**) shows that dense NMDAR1 immunoreactivity overlaps with PV-ir dendritic bundles. NMDAR1 immunoreactivity is highest in layer 2. *Arrowheads* point to corresponding regions. *Bar* 200 μm

with a particular involvement in plasticity, such as might be expected for a region closely linked with the hippocampal memory system.

We identified, as another indicator of the complex organization of the superficial layers in the GRS, a separate, honeycomb-like mosaic at the border of layers 1 and 2. This consists of hollows surrounded by walls containing both dendrite- and terminal-like PV-ir profiles. The center-to-center spacing of the hollows is 58–112 μm (mean 80 μm). The honeycomb structure is continuous with adjacent neocortical areas, including visual areas and anterior cingulate cortex.

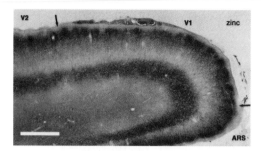

FIG. 2. Coronal section of the caudal part of V1, stained with zinc histochemistry. A periodic pattern is observed in the superficial layers. *Arrows* demarcate medial and lateral borders of V1. *ARS*, agranular retrosplenial cortex. *Bar* 500 μm

Rat Visual Cortex

In the rat, visual cortex is less specialized than the somatosensory barrel cortex in the same species or the visual cortex in primates. This pertains to its physiological organization and the manner of thalamocortical terminations in layer 4. V1 also has substantial connections with nonvisual areas, including retrosplenial cortex [9].

In rat visual cortex, as in GRS, the walls of the honeycomb contain a higher density of PV-ir neuropil. The pattern can be seen in coronal section, where the hollows appear as a series of notches. It is more easily demonstrated by using tangential sections and double labeling to reveal overlapping or complementary patterns. Zinc-rich corticocortical terminals (Fig. 2), GluR2/3, NMDAR1, and calbindin (CB) also colocalize in the honeycomb walls. Other elements, notably GABAaα1 and thalamocortical terminals, visualized by antibody against vesicular glutamate transporter 2 [10] and cytochrome oxidase (CO), are preferentially dense in the hollows [6]. The complementary distribution of zinc and CO recalls previous reports of an interdigitated columnar mosaic in kittens or primates, although this was in layers 3–4 in these species [11]. In V1, the center-to-center spacing of the hollows is about 80 μm (range: 50–120 μm). This small-scale modularity is also similar to the layer 4A honeycomb in primate area V1, but there are otherwise some differences. Notably, in layer 4A, thalamic afferents coincide with PV-ir neuropil in the walls [12].

The dendritic components of the honeycomb are somewhat less clear in V1 than in the GRS. This is in part because of the technical difficulty of selectively labeling specific subpopulations, especially the thinner dendrites of layer 2 pyramidal neurons. However, MAP2 staining reveals bundles of thick apical dendrites originating from pyramidal neurons in layers 3 and 5 [12]. These tend to occupy the PV-ir hollows, overlapping with thalamocortical terminations. Thinner dendrites, probably from more superficial neurons, are more concentrated in the PV-ir walls.

Conclusion

Visual cortex is increasingly considered as interactive, rather than a quasi-isolated vision-dedicated module. From this perspective, it may be appropriate to reevaluate the implied preeminent role of thalamocortical connections to layer 4. Here, we have reviewed evidence for a pronounced honeycomb mosaic at the border of layers 1 and 2 in rat visual cortex. This appears to be continuous with a similar structure in GRS. Both may be associated with multiple parallel circuits, possibly related to plasticity.

References

1. Churchland PS, Ramachandran VS, Sejnowski TJ (1994) A critique of pure vision. In: Koch C, Davis JL (eds) Large-scale neuronal theories of the brain. MIT Press, Cambridge, pp 23–60
2. Casagrande VA, Xu X, Sary G (2002) Static and dynamic views of visual cortical organization. In: Azmitia E et al. (eds) Changing views of Cajal's neuron. Elsevier, Amsterdam, pp 389–408
3. Paradiso MA (2002) Perceptual and neuronal correspondence in primary visual cortex. Curr Opin Neurobiol 12:155–161
4. Falchier A, Clavagnier S, Barone P, et al. (2002) Anatomical evidence of multimodal integration in primate striate cortex. J Neurosci 22:5749–5759
5. Rockland KS, Ojima H (2001) Calcarine area V1 as a multimodal convergence area. Neurosci Abstr 27
6. Ichinohe N, Rockland KS (2002) Honeycomb-like structure at the border of layers 1 and 2 in the cerebral cortex. Jp Neuroscience Abstr 25
7. Ichinohe N, Rockland KS (2002) Parvalbumin positive dendrites co-localize with apical dendritic bundles in rat retrosplenial cortex. Neuroreport 13:757–761
8. Wyss JM, van Groen T, Sripanidkulchai K (1990) Dendritic bundling in layer I of granular retrosplenial cortex: intracellular labeling and selectivity of innervation. J Comp Neurol 295:33–42
9. Softon AJ, Dreher B (1995) Visual system. In: Paxinos G (ed) The rat nervous system. Academic Press, San Diego, pp 833–880
10. Fujiyama F, Furuta T, Kaneko T (2001) Immunocytochemical localization of candidates for vesicular glutamate transporters in the rat cerebral cortex. J Comp Neurol 435:379–387
11. Dyck RH, Cynader MS (1993) An interdigitated columnar mosaic of cytochrome oxidase, zinc, and neurotransmitter-related molecules in cat and monkey visual cortex. Proc Natl Acad Sci USA 90:9066–9069
12. Peters A, Sethares C (1991) Organization of pyramidal neurons in area 17 of monkey visual cortex. J Comp Neurol 306:1–23

Cross-Correlation Study of Network Dynamics in the Cat Primary Visual Cortex

Tomoyuki Naito, Hirofumi Ozeki, Osamu Sadakane, Takafumi Akasaki, and Hiromichi Sato

Key words. Primary visual cortex, Cross-correlograms, Classical receptive field

To assess the network dynamics in the primary visual cortex, we performed simultaneous recordings of neuronal activity from cell pairs within 1 mm apart in anesthetized and paralyzed cats. Cross-correlograms (CCs) were calculated for the responses to grating stimuli with varying orientations and sizes for the 30 cell pairs to date. There were three patterns of cell pairs in respect of tuning property to stimulus sizes. First, both cells exhibited strong suppression of firings with stimuli larger than their classical receptive field (CRF). Second, only one cell exhibited suppression to the large stimuli while the other did not. In these two patterns, the contribution of the peak CCs became smaller when the grating stimulus was larger than the CRFs. And third, both cells did not significantly reduce activity to the grating patch larger than their CRFs. In this case, the peak contribution in CCs did not vary with an increase of the stimulus size. We concluded that the strength of firing interaction is positively correlated with the change in firing rate of cell pairs because of the change of stimulus sizes. We also analyzed CCs in respect of orientation preference of cell pairs. Although the significant peak CCs were most often observed for the common and optimal orientation of the cell pairs, there were some pairs that showed strong peaks for the nonpreferred orientation (more than 60° apart from the optimal) of stimulus, which evoked responses with low firing levels. This result suggests that the strength of firing correlation of cell pairs does not simply depend on the firing rate when the stimulus orientation is varied.

School of Health and Sport Sciences, Osaka University, 1-17 Machikaneyama, Toyonaka, Osaka 560-0043, Japan

Contrast-Dependent Gain Control of the Contextual Response Modulation in the Cat Primary Visual Cortex

Osamu Sadakane, Hirofumi Ozeki, Tomoyuki Naito, Takafumi Akasaki, and Hiromichi Sato

Key words. Receptive field, Surround suppression, Contextual modulation, Contrast, V1

Introduction

Neurons of the primary visual cortex (V1) respond to stimuli presented at the classical receptive field (CRF), and the responses are suppressively modulated by stimuli simultaneously presented at the receptive field surround (SRF) [1, 2]. The response modulation by an SRF stimulus depends on the stimulus orientation and spatial frequency; thus, it was suggested that intracortical mechanisms contribute to this phenomenon [2]. We recently reported that this suppressive modulation is not due to a direct inhibition to the cell by the intracortical inhibition [3] (also see the chapter by H. Ozeki et al., this volume).

There are some studies showing that the direction of the modulatory effect of an SRF stimulus changes depending on the stimulus contrast [4, 5]. For example, Sengpiel et al. [4] reported that the effect of an SRF stimulus is suppressive when the CRF stimulus contrast is high and facilitatory when it is low.

In the present study, we recorded responses of cat V1 neurons in order to assess the contrast dependency of the surround effect on responses to CRF stimulation with either a high- or low-contrast grating. Our working hypothesis is that excitatory inputs evoked by the stimulation of CRF and SRF are summative, and when total excitatory inputs exceed a certain threshold level, the SRF effect changes from facilitatory to suppressive. To test this hypothesis, we examined the SRF effect at different saturation levels of CRF responses.

Materials and Methods

Details of the procedure have been described previously [1]. Adult cats were anesthetized with fentanyl citrate (Fentanest, Sankyo, Tokyo, Japan; 10 µg/kg/h, i.v.) and isoflurane [0%–0.5% in N_2–O_2 (2:1)], paralyzed with pancuronium bromide

School of Health and Sport Sciences, Osaka University, 1-17 Machikaneyama, Toyonaka, Osaka 560-0043, Japan

(0.1 mg/kg/h, i.v.), and maintained under artificial ventilation. Visual stimuli of drifting sinusoidal gratings were generated with VSG2/3 (Cambridge Research Systems, UK) and presented on a cathode ray tube display (mean luminance $40 \, cd/m^2$, screen size $40 \times 30 \, cm^2$), which was placed 57 cm in front of the animals. Glass micropipettes filled with 0.5 M sodium acetate containing 4% Pontamine Sky Blue for marking were used to record single-cell activities extracellularly. When a cell was encountered, the receptive field was manually plotted, and then a circular grating stimulus was targeted to the CRF center. For the stimulation of CRF and SRF, the stimulus orientation and spatial frequency were optimal for the cell. The size of the CRF stimulus was adjusted to the peak of the stimulus size tuning curve. The SRF stimulus was presented on the entire screen. The contrast-response relationships were fitted by

$$R(c) = R\mathrm{max}\frac{c^n}{c^n + \sigma^n} + b$$

where c is the stimulus contrast, R the cell's response, b the spontaneous activity, and $R\mathrm{max}$, σ, and n the free parameters.

Results

Forty-five cells exhibiting suppressive modulation by the SRF stimulation were analyzed. Our working hypothesis predicts that a high-contrast visual stimulation of the CRF will make the firing rate of the cell close to the saturation level at which the suppressive surround effect is induced. When the CRF stimulus contrast is low, however, the cell has more room to increase its activity; then the SRF stimulation results in greater excitation, which makes the effect facilitatory.

To examine these points, first, we measured the contrast-response relationship for a CRF stimulus. Then, the CRF stimulus contrast was fixed at either a high or a low value, and the SRF stimulus contrast was varied. Figure 1a shows the contrast-response curve for the CRF response of a cell recorded from layer 4. The CRF stimulus contrast was varied, while other parameters were fixed to be optimal. Dashed lines indicate the contrasts we used in the contrast tuning test of the surround effect shown in Fig. 1b. At 2% CRF stimulus contrast, the cell responded with only a negligible number of spikes, and at 30% CRF stimulus contrast, the response reached the saturation level. Figure 1b shows the SRF-contrast-response curves when the SRF stimulus contrast was varied, while the CRF stimulus contrast was fixed at 2%, 10%, or 30%. With an increment of the contrast of the SRF grating, the SRF grating progressively suppressed the CRF responses regardless of the contrast of the CRF grating. Even when the CRF stimulus contrast was 2%, no facilitatory surround effect was observed. In the 45 cells analyzed, the modulatory effect of SRF gratings was always suppressive, and this effect became stronger as the SRF grating contrasts increased.

Figure 2 shows the comparison of the suppressive effect of a SRF stimulus on the responses to high- and low-contrast CRF stimulus. To evaluate the suppressive SRF effect, we calculated the value of SRF stimulus contrast (SC50) at which 50% suppression of the maximal suppression is induced. In this graph, SC50 values of two (high and low) CRF stimulus contrasts are plotted for each neuron ($N = 45$; abscissa,

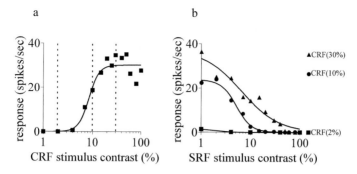

FIG. 1a,b. An example of the contrast-tuning curve of response to classical receptive field (*CRF*) stimulation (**a**) and those to the combined stimulation of the CRF and the receptive field surround (*SRF*; **b**). This cell was recorded from layer 4. **a** Data points were fitted with a hyperbolic function. *Dashed lines* indicate the contrast values (2%, 10%, 30%) used in the test shown in **b**. **b** Effects of SRF stimulus contrast on the cell's response to CRF stimulation at three different contrast levels. *Symbols on the right side* of the graph indicate activity level in response to the CRF stimulus alone at three different contrast levels

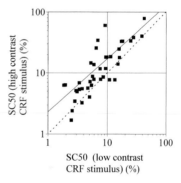

FIG. 2. Comparison of suppressive effects of SRF stimulus on responses of 45 cells to high and low contrast CRF stimulus. The values of SRF stimulus contrasts at one-half the strength of suppressive effects (*SC50*) were plotted for the responses to high- and low-contrast CRF stimulus. The *dashed line* and the *solid line* indicate the diagonal line and the regression line on log–log axes ($y = 0.87x + 0.34$), respectively

SC50 for low CRF stimulus contrast; ordinate, SC50 for high CRF stimulus contrast). Most plots fell above the diagonal line (dashed line); that is, SC50 values for high CRF stimulus contrast were larger than those for low CRF stimulus contrast, suggesting the presence of a contrast-gain control mechanism.

Discussion

In the present study, when a wide area ($40° \times 30°$) surrounding the CRF was stimulated with a drifting uniform grating, the response to the CRF stimulation was progressively suppressed with an increment of the surround contrast regardless of the CRF stimulus contrast. Basically, the tuning curves for the surround contrast at both high and low CRF stimulus contrast conditions showed profiles that were quite similar to each other, suggesting the contrast-invariant property of the surround suppression.

These results are against our working hypothesis that there is a switching mechanism for the additivity of the CRF- and SRF-derived excitations between facilitatory and suppressive effects according to the input level. Thus, the effect of the stimulation of the receptive field surround is gain-controlled in order to adjust the magnitude of response modulation according to the contrast level.

Our results are contradictory to those of previous studies that reported the facilitatory surround effect, particularly when the CRF stimulus contrast was low [4, 5]. Concerning this point, Sceniak and colleagues [6] reported that the response summation area of V1 neurons depends on the grating contrast and is on average 2.3-fold greater at low contrast than that at high contrast. Our preliminary results suggest that the annular area between two CRFs of different diameters determined by the high- and low-contrast gratings is responsible for the facilitatory surround effect. That is, when the low-contrast grating was presented at the CRF with a diameter determined by the high-contrast grating, the effect of stimulating the annular area with the low-contrast annular grating was facilitatory, whereas when the high-contrast annular grating was presented, the effect was suppressive regardless of its contrast (data not shown). However, grating stimuli larger than the CRF determined by the low-contrast grating always suppressed the response regardless of the CRF contrast. This can explain why researchers using a small display for the visual stimulation often reported the facilitatory surround effect, whereas we stimulated the surround area (screen size, $40° \times 30°$) that is much larger than the CRF for the low-contrast stimulus and rarely observed the facilitatory effect.

Acknowledgments. This work was supported by grants from MEXT (13041032, 13210086, 14017058, NRV project) and the Naito Foundation to H.S. H.O. was supported by the JSPS Research Fellowship.

References

1. Akasaki T, Sato H, Yoshimura Y, et al. (2002) Suppressive effects of receptive field surround on neuronal activity in the cat primary visual cortex. Neurosci Res 43:207–220
2. DeAngelis GC, Freeman RD, Ohzawa I (1994) Length and width tuning of neurons in the cat's primary visual cortex. J Neurophysiol 71:347–374
3. Ozeki H, Akasaki T, Sadakane O, et al. (2001) Possible mechanism underlying the contextual response modulation in the cat primary visual cortex. Abstr Soc Neurosci 27:619–622
4. Sengpiel F, Sen A, Blakemore C (1997) Characteristics of surround inhibition in cat area 17. Exp Brain Res 116:216–228
5. Polat U, Mizobe K, Pettet MW, et al. (1998) Collinear stimuli regulate visual responses depending on cell's contrast threshold. Nature 391:580–584
6. Sceniak MP, Ringach DL, Hawken MJ, et al. (1999) Contrast's effect on spatial summation by macaque V1 neurons. Nat Neurosci 2:733–739

Contribution of Excitation and Inhibition to the Stimulus Size- and Orientation-Tuning of Response Modulation by the Receptive Field Surround in the Cat Primary Visual Cortex

Hirofumi Ozeki, Osamu Sadakane, Takafumi Akasaki, Tomoyuki Naito, and Hiromichi Sato

Key words. Receptive field surround, Contextual modulation, Intracortical inhibition, Bicuculline iontophoresis, V1

Introduction

The response of neurons in the primary visual cortex (V1) is modulated by the stimulus presented at the surround of the classical receptive field (CRF), and the predominant effect of this surround stimulation is suppressive [1–5]. The strength of the surround suppression is dependent on the relationship of stimulus parameters, such as orientation and spatial frequency, between grating stimuli at the inside and outside of the CRF [1, 2]. Although properties of the surround suppression have been well characterized, input mechanisms underlying the stimulus-context-dependent response modulation and the size-tuning properties of V1 neurons have not been clarified yet. There is evidence that the surround suppression in V1 is both subcortical [5] and intracortical [2, 4] in origin.

In the present study, we investigated the possible contribution of the intracortical inhibition to the surround suppression by extracellular recordings of responses of cat V1 neurons. We examined the effects of blocking intracortical inhibition with an iontophoretic administration of bicuculline methiodide (BMI), a selective antagonist of γ-aminobutyric acid $(GABA)_A$ receptors, on the size-tuning properties and the orientation dependency of the surround effect.

Materials and Methods

Details of procedures have been described previously [1]. Adult cats were anesthetized with fentanyl citrate (Fentanest, Sankyo, Tokyo, 10 μg/kg/h, i.v.) and isoflurane [0%–0.5% in N_2O–O_2 (2:1)], paralyzed with pancuronium bromide (0.1 mg/kg/h, i.v.),

School of Health and Sport Sciences, Osaka University, 1-17 Machikaneyama, Toyonaka, Osaka 560-0043, Japan

and maintained under artificial ventilation. Visual stimuli of drifting sinusoidal gratings were generated with VSG2/3 (Cambridge Research Systems, Rochester, UK) and presented on the cathode ray tube display (mean luminance, 40 cd/m²; screen size, 40 × 30 cm²), which was placed 57 cm in front of the animal. When a cell was encountered, the minimum response field of the cell was manually plotted. Then a circular grating stimulus was targeted at the center of the field. Each stimulus was presented stationary for 1 s, and then drifted for 2–4 s. This was repeated 2–20 times in pseudo-random order to accumulate peristimulus time histograms with an interleave of 1–4 s. Three-barreled glass micropipettes were used for extracellular single-unit recordings and iontophoretic administration of BMI (Sigma, St. Louis, USA, 5 mM, pH 4.0) and GABA (Tocris, Buckhurst Hill, UK, 0.5 M, pH 4.0) [6]. The tip of the recording electrode protruded by 20–30 μm from those of the drug pipettes. Ejection currents of BMI were determined so as to antagonize the effect of iontophoretically administered GABA, which was sufficient to abolish visual responses of the observed cell. Usually, ejecting and retaining currents of BMI were between +5 and +20 nA and between −5 and −10 nA, respectively. The recording pipettes were filled with 0.5 M sodium acetate containing 4% Pontamine Sky Blue for marking. The spike response evoked during the first 2 s to drifting grating was analyzed to compute the mean firing rate (F0) and first harmonic component (F1) of the response, and the F0 and F1 were used as the response measures for complex and simple cells, respectively.

Results

Figure 1a shows an example of the stimulus size tuning of responses of the cell, which exhibited the suppressive response modulation by stimulation of the CRF surround. The size of the grating stimulus was varied, while the other grating parameters were optimal for the cell. In the control (solid line), the response peaked at 1.4° of the

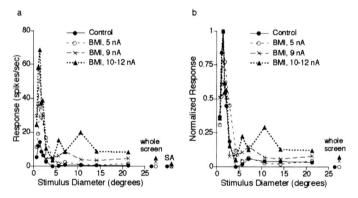

FIG. 1a,b. Effects of bicuculline methiodide (*BMI*) on the stimulus size tuning of a layer 2/3 simple cell. **a** Tuning curves for the control response and responses during the BMI administration with different ejecting currents. **b** Tuning curves normalized to the maximal responses. *Whole screen*, response to whole screen stimulation (40° × 30°); *SA*, spontaneous activity

stimulus diameter (the CRF size), and then decreased at the stimulus size larger than 1.4° in diameter. During the BMI administration with +5, +9, and +10–12 nA of ejecting current (dashed lines), the response to the CRF stimulus was remarkably facilitated depending on the current intensity. However, the responses to stimuli larger than 4.3° in diameter were not noticeably facilitated. Figure 1b shows the normalized stimulus size-tuning curves of the responses. The strength of the maximal surround suppression was calculated as "peak amplitude minus maximally suppressed amplitude" in the normalized tuning curves. Spontaneous activity was negligible both in the control and during the BMI administration. Even though there were 0.1%, 5.3%, and 12.9% reductions of the surround suppression for +5, +9, and +10–12 nA of BMI administration, respectively, the surround suppression was not abolished by the blockade of the intracortical inhibition. As an average of all 43 cells recorded, a 12.8% reduction of the surround suppression was observed (mean BMI current, 14 ± 11 nA), and there was no difference of the BMI effects between simple ($N = 22$) and complex ($N = 21$) cells. This result suggests that the surround suppression is not due to a predominant operation of the intracortical inhibition. To elucidate the contribution of excitation and inhibition to the size-tuning property, we compared the tuning curve during the BMI administration (excitation) with that of the control (excitation and inhibition), and estimated the relative contribution of inhibition (response with BMI minus that of the control). As a result of this analysis, we found that the excitation and inhibition have size-tuning properties similar to each other, and this suggests that both inputs are well balanced throughout a wide range of the stimulus sizes.

Figure 2a shows an example of the orientation tuning of the suppressive surround modulation in a complex cell. In this test, while presenting the drifting grating with optimal parameters at the CRF, the surround of the CRF was stimulated by the drifting grating with the same parameters as the CRF grating but with varying orientations. In the control (solid line), the response to the CRF stimulation was more strongly suppressed when the orientation contrast between the center and surround

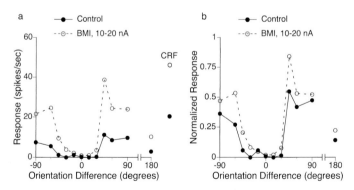

FIG. 2a,b. Effects of BMI on the orientation-contrast dependency of the surround effect in a complex cell recorded from a deep layer. **a** Tuning curves for the control response and response with BMI administration. *Abscissa* (orientation difference), difference of orientation between the classical receptive field (*CRF*) and surround gratings. **b** Tuning curves normalized to the maximal responses to the CRF stimulation

gratings was small. Although the BMI administration with +10–20 nA facilitated the responses, the orientation dependency of the surround effect was not affected (dashed line), as shown in Fig. 2b (the tuning curves normalized to the responses to the CRF stimulation). This result suggests that the orientation contrast dependency of the surround suppression is not due to a predominant operation of the intracortical inhibition. Similar to the analysis for the stimulus size tuning, we estimated the contribution of excitation and inhibition, and found that both inputs are well balanced throughout the wide range of the surround orientations.

Discussion

Cortical mechanisms underlying the size tuning and the orientationally tuned surround suppression of V1 neurons were studied with the pharmacological blockade of the intracortical inhibition. First, since the BMI administration did not significantly affect the tuning properties of the surround suppression, and the estimated excitation and inhibition exhibited similar tuning profiles, it is suggested that the well-balanced excitation and inhibition cooperatively control the response gain of V1 neurons. Second, our results contradict the suggestion of previous studies that the suppressive response modulation is derived from the intracortical inhibition [2–4]. These results suggest that the primary cause of the cortical surround suppression is a reduction of afferent inputs to V1, probably due to the length-tuning property of the lateral geniculate nucleus [5]. However, we point out that the excitatory and inhibitory size tuning exhibited profiles similar to each other, which could be a basis for the input-level-invariant response modulation by the receptive field surround as well as the stimulus specificity of the single cell responses [6].

Acknowledgments. This work was supported by grants from MEXT (13041032, 13210086, 14017058, NRV project) and the Naito Foundation to H.S. H.O. was supported by the JSPS Research Fellowship.

References

1. Akasaki T, Sato H, Yoshimura Y, et al. (2002) Suppressive effects of receptive field surround on neuronal activity in the cat primary visual cortex. Neurosci Res 43:207–220
2. DeAngelis GC, Freeman RD, Ohzawa I (1994) Length and width tuning of neurons in the cat's primary visual cortex. J Neurophysiol 71:347–374
3. Walker GA, Ohzawa I, Freeman RD (1999) Asymmetric suppression outside the classical receptive field of the visual cortex. J Neurosci 19:10536–10553
4. Anderson JS, Lampl I, Gillespie DC, et al. (2001) Membrane potential and conductance changes underlying length tuning of cells in cat primary visual cortex. J Neurosci 21:2104–2112
5. Jones HE, Andolina IM, Oakely NM, et al. (2000) Spatial summation in lateral geniculate nucleus and visual cortex. Exp Brain Res 135:279–284
6. Sato H, Katsuyama N, Tamura H, et al. (1996) Mechanisms underlying orientation selectivity of neurons in the primary visual cortex of the macaque. J Physiol 494:757–771

Effects of Chronic Exposure to Vertical Orientation on the Development of Orientation Preference Maps

Shigeru Tanaka, Jérôme Ribot, and Kazuyuki Imamura

Key words. Postnatal development, Visual experience, Orientation preference, Intrinsic optical imaging

Introduction

How the functional maps develop during the critical period of the visual cortex is one of the major questions that have challenged many experimental and theoretical investigators. Although it is accepted that visual experience contributes to the maturation of ocular dominance and orientation maps, how and to what extent visual experience affects map formation has not been fully elucidated yet. Blakemore and Cooper [1] have reported that neurons in area 17 exhibited orientation polar histograms strongly biased to the experienced orientation in cats that had been exposed to only vertical or horizontal stripes in a cylindrical room for several hours a day during development. Although such a strong bias of preferred orientations could not be reproduced by other groups following almost the same experimental protocol [2, 3], many investigators were motivated to investigate the role of visual experience for orientation map formation. Recently, Sengpiel et al. [4] have shown by intrinsic optical recording that the basic structure of orientation maps is rather robust even when the experienced orientation is slightly overrepresented in the visual cortex. It is thought that the discrepancy between the results is partly due to the difficulty in achieving perfect restriction of stimulus orientations for a long time. In this study, we mounted chronically special goggles made of cylindrical lenses on kittens to restrict their exposure to a single orientation during development. Only vertically elongated images of the environment could be seen through the goggles. We recorded intrinsic optical signals from the visual cortex of these kittens, and demonstrated that vertical orientation was strongly overrepresented in all the animals examined. The degree of overrepresentation for the goggles-mounted kittens was much stronger than that for a kitten exposed to vertical stripes inside a cylindrical room.

Laboratory for Visual Neurocomputing, Brain Science Institute, RIKEN, 2-1 Hirosawa, Wako, Saitama 351-0198, Japan

Materials and Methods

We used five young kittens delivered from our own colony. One of the kittens examined (Kitten A) was normally reared in the colony and used as a control. Another kitten (Kitten B) was exposed to vertically oriented black-and-white stripes inside a cylindrical room according to Blakemore and Cooper's experimental protocol [1]. In our study, the kitten stayed inside the cylindrical room for about 2 h a day from postnatal day 2 to day 55. At other times, it was kept in a dark room. During the exposure to vertical stripes, we manipulated the length of the leash on the animal so that the animal was able to walk around and sit down but could not lie down. For the other kittens, we mounted goggles of acrylic cylindrical lenses on postnatal day 24 (Kitten C), day 19 (Kitten D), and day 31 (Kitten E). The frame of the goggles was connected to a metal slab and then the slab was fixed with a head holder anchored to the animals' forehead bone. The goggles were removable for cleaning. They were sometimes replaced with goggles of different sizes according to the growth of the animal's head. We adjusted the position of the goggles so that the animals would see vertically elongated images of their environment in the lower visual field that is retinotopically represented in the optically recorded region.

Intrinsic optical recordings were conducted using five kittens between the ages of 33 and 77 days. All the kittens were anesthetized and paralyzed during the recording. The cortex was illuminated with light of 700-nm wavelength. Images of intrinsic optical signals were captured with a CCD video camera. The animals were stimulated with full-screen moving square-wave gratings with a spatial frequency of 0.15 or 0.5 cycle/degree in six orientations with the same interval of orientation angles (30°).

Results

Orientation polar maps, orientation angle maps, and orientation histograms for three kittens are shown in Fig. 1. In the normal kitten, the orientation polar and angle maps, respectively, revealed continuous and periodic arrangements of preferred orientations except at the pinwheel centers (A1 and A2 in Fig. 1). The distribution of preferred orientations was almost flat (A3). For the cylindrical-room-reared kitten (Kitten B), the angle map (B2) showed that the basic structure of the orientation arrangement was not very different from that for the normal kitten (A2). However, in the polar map (B1), iso-orientation domains representing the vertical orientation (yellow domains) showed a much higher magnitude of orientation preference. The orientation distribution of this kitten has a peak around the vertical orientation (B3), indicating that the experienced orientation was slightly overrepresented, which agrees with Sengpiel et al.'s results [4]. In contrast, for all goggles-mounted kittens (Kittens C, D, and E) the periodic arrangement of preferred orientations was at least partly disrupted. In their polar maps (C1), the yellow domains stood out, indicating that domains eliciting strong responses agree well with the iso-orientation domains representing vertical orientation. In both the polar and angle maps (C1 and C2), overrepresentation of the experienced orientation was evident. Indeed, the orientation distribution (C3) exhibited a sharp peak at the vertical orientation, and the range of the overrepresented orientations was narrower than that for the cylindrical-room kitten. Such extremely

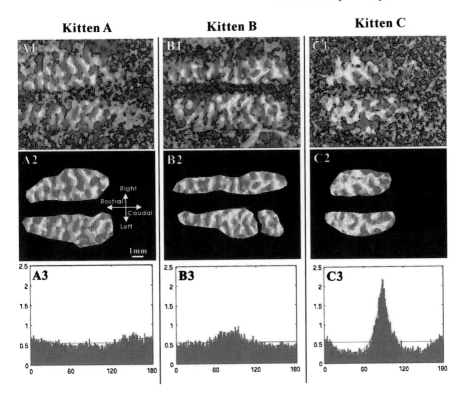

Fig. 1A–C. Orientation polar maps (A1–C1), angle maps inside strongly responsive domains (A2–C2), and orientation distributions (the number of pixels plotted against preferred orientation) (A3–C3) are shown for normal (Kitten A), cylinder-room (Kitten B), and goggles-mounted (Kitten C) kittens

strong overrepresentation of the vertical orientation was also observed for the other two goggles-mounted kittens (Kittens D and E). The difference in overrepresentation width suggests that constraint on the experienced orientation for the goggles-mounted kittens was tighter than that for the cylindrical-room kitten. The 2-week exposure to vertical orientation was sufficient for achieving the strong overrepresentation.

Discussion and Conclusion

To elucidate the effects of visual experience on orientation map formation during the critical period of the visual cortex, we reared kittens in a very restricted orientation experience by means of chronically mountable goggles and observed orientation maps during their development using the intrinsic optical imaging technique. The strict constraint in exposure to vertical orientation gave rise to extremely strong overrepresentation of the vertical orientation at the cost of the disruption of the basic structure of the orientation maps. The overrepresentation effect could be recognized in both the domain size and response strength. Sengpiel et al. [4] used cylindrical lens goggles combined with the cylindrical-room training to impose a strong restriction on experienced orientations. However, the experienced orientation was slightly over-represented in the visual cortex, whereas the basic structure of the orientation maps was retained. Therefore, they concluded that orientation map formation is robust. What are the origins of the discrepancy between our experimental finding and theirs in term of the degree of overrepresentation? Our goggles-mounted kittens were exposed to vertical orientation whenever they were awake. On the other hand, the cylindrical-room kittens experienced vertical orientation only for a few hours each day. In addition, in the cylindrical-room protocol, the kittens stayed in darkness most of time [1–4]. During dark rearing, some intrinsic mechanisms such as correlation-based learning [5, 6] may contribute to orientation map formation and promote normal orientation preferences even if the driving force of map formation is weak. Since the intrinsic mechanisms are independent of visual experience, the orientation map may recover from the overrepresentation resulting from the restricted orientation experience inside the cylindrical room. Thus, it is suggested that dark rearing caused the weak overrepresentation of the experienced orientation in the cylindrical-room kittens compared with that of the goggles-mounted kittens.

Acknowledgments. We thank Yuji Akimoto and Katsuya Ozawa for their assistance with the experiments and for maintaining all the kittens examined.

References

1. Blakemore C, Cooper GF (1970) Development of the brain depends on the visual environment. Nature 228:477–478
2. Stryker MP, Sherk H (1975) Modification of cortical selectivity in the cat by restricted visual experience: a reexamination. Science 190:904–906
3. Rauschecker JP, Singer W (1981) The effects of early visual experience on the cat's visual cortex and their possible explanation by Hebb synapses. J Physiol 310:215–239
4. Sengpiel F, Stawinski P, Bonhoeffer T (1999) Influence of experience on orientation maps in cat visual cortex. Nat Neurosci 2:727–732
5. Miller KD (1994) A model for the development of simple cell receptive fields and the ordered arrangement of orientation columns through activity-dependent competition between ON- and OFF-center inputs. J Neurosci 14: 409–441
6. Miyashita M, Kim DS, Tanaka S (1997) Cortical directional selectivity without directional experience. Neuroreport 8:1187–1191

Self-Organization Model of Ocular Dominance Columns and Cytochrome Oxidase Blobs in the Primary Visual Cortex of Monkeys and Cats

Hayato Nakagama and Shigeru Tanaka

Key words. Visual cortex, CO blobs, OD columns, Self-organization model, Development

Introduction

In the primary visual cortex of monkeys and cats, regions relatively strongly stained by cytochrome oxidase (CO) appear to be patches arranged over the cortical surface, which are called CO blobs. In macaques or galagos, CO blobs have been found along the middle of ocular dominance (OD) bands [1], which indicates that CO blobs and OD columns do not develop independently, while the structural relationship between CO blobs and OD columns in cats [2] or squirrel monkeys [3] is not clear. Such observation raises a question of what mechanism is commonly underlying the formation of CO blobs and that of OD columns. Here, we propose a self-organization model based on the Hebbian learning rule that accounts for the coordinated development of CO blobs and OD maps observed in monkeys and cats.

Materials and Methods

Four types of inputs are assumed to be terminated at layers 2/3 in the visual cortex (Fig. 1). Two of them receive inputs from the left eye and the other two receive inputs from the right eye. The ocularity is represented by μ_1: $\mu_1 = L$ for the left-eye inputs and $\mu_1 = R$ for the right-eye inputs. These types of inputs are subdivided into groups A and B, which are represented by μ_2: $\mu_2 = A$ for group A and $\mu_2 = B$ for group B. The positions in the visual field of projection neurons and the cortical position of afferent input terminals are indicated by k and j, respectively. The synaptic connection from the projection neuron specified by (k, μ_1, μ_2) to the cortical position j is represented by σ_{j,k,μ_1,μ_2}, which takes a value of 1 when being connected, and 0 when being disconnected. Each cortical position receives a single synaptic input due to the

Laboratory for Visual Neurocomputing, Brain Science Institute, RIKEN, 2-1 Hirosawa, Wako, Saitama 351-0198, Japan

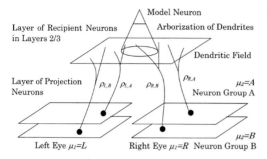

FIG. 1. Structure of the model

winner-take-all mechanism based on the competition among afferent inputs at the same dendritic spine [4].

The model is described by the following energy function to minimize:

$$H = -\sum_{j,j'} \sum_{k,\mu_1,\mu_2} \sum_{k',\mu_1',\mu_2'} V_{j,j'} \Gamma_{k,\mu_1,\mu_2,k',\mu_1',\mu_2'} \sigma_{j,k,\mu_1,\mu_2} \sigma_{j',k',\mu_1',\mu_2'}$$

$$+ \sum_{k,\mu_1,\mu_2} c_{k,\mu_1,\mu_2} \left(\sum_j \sigma_{j,k,\mu_1,\mu_2} \right)^2. \tag{1}$$

Here, $V_{jj'}$ represents the cortical interaction between afferent inputs at positions j and j', which is given by the Gaussian function $\Gamma_{k,\mu_1,\mu_2;k',\mu_1',\mu_2'}$, in the energy function represents the correlation of firings between a pair of neurons specified by (k, μ_1, μ_2) and (k', μ_1', μ_2'), and is given by

$$\Gamma_{k,\mu_1,\mu_2;k',\mu_1',\mu_2'} = \gamma^{\frac{1}{2}}_{\mu_1,\mu_2} \gamma^{\frac{1}{2}}_{\mu_1',\mu_2'} \left((1-h)\delta_{\mu_1,\mu_1'} + h \right) \left((1-g)\delta_{\mu_2,\mu_2'} + g \right) \sum_l R_{k;l} R_{k';l}. \tag{2}$$

Here, h represents the strength of correlation in activities between inputs from the left and right eyes in the same-group input pathway, and g represents the strength of correlation in activities between inputs from groups A and B in the same-eye input pathway. $\delta_{\mu_1,\mu_1'}$ and $\delta_{\mu_2,\mu_2'}$ are Kronecker's deltas, which take a value of 1 for $\mu_i = \mu_i'$ and 0 for $\mu_i \neq \mu_i'$ ($i = 1$ or 2). γ_{μ_1,μ_2} represents the average firing rate conveyed by inputs of μ_1 and μ_2. Sequences of neuronal spikes can be described approximately by the Poisson process; hence, the standard deviation of the activity is proportional to the square root of the mean firing rate, $\gamma^{1/2}_{\mu_1,\mu_2}$. Therefore the correlation function $\Gamma_{k,\mu_1,\mu_2;k',\mu_1',\mu_2'}$ is scaled by the square root of the mean firing rates of inputs $\gamma^{1/2}_{\mu_1,\mu_2}$ [5]. R_{kl} represents the normalized activity of a lateral geniculate nucleus (LGN) cell at position k in response to the stimulus at the position l in the visual field, which is given by the Gaussian function. The second term of the energy function represents the constraint on the number of active synaptic inputs originating from individual projection neurons so that the total number of active synaptic inputs from each projection neuron $\sum_j \sigma_{j,k,\mu_1,\mu_2}$ tends to be as small as possible. c_{k,μ_1,μ_2} is a certain positive constant that determines the magnitude of the constraint.

An array of 64 × 64 cells is put on each layer of neurons that send afferent inputs to the simulated cortex. The recipient layer, which is assumed to be the superficial layer in the cortex, is composed of 256 × 256 grids and receives afferent inputs from neurons in the projection layers. The periodic boundary condition is adopted to avoid finite-size effect as much as possible. At the beginning of the simulation, each cortical position randomly receives a synaptic input from a projection neuron. In the simulation, for each cortical position, a new candidate afferent input is selected randomly, and the present input is replaced with a new one according to Eq. 1 by using the simulated annealing methods [5].

Results

Figure 2 shows the patterns of afferent inputs after self-organization. Dark and light pixels represent synaptic terminals of left- and right-eye-specific inputs, respectively. Moreover, blob-like structure can be seen in which the blobs are placed in the middle of the OD bands in Fig. 2A, like the CO blobs observed in the primary visual cortex of monkeys [1]. In order for synaptic terminals from groups A and B to form a blob structure, imbalance in activities between groups A and B is needed ($\gamma_{L,A} = \gamma_{R,A} = 2.5$, $\gamma_{L,B} = \gamma_{R,B} = 0.4$). This is consistent with the fact that neurons in the CO blobs are metabolically more active than neurons outside the blobs [6].

In monkeys, OD columns [7] and CO blobs [8] are observed at birth, and are therefore formed before the experience of pattern vision. It is thus expected that these structures in monkeys can be formed in the absence of correlation in between-eye activities. When the correlation strength between the two eyes h is set to be 0.0 in the simulation, blobs are placed in the middle of OD bands (Fig. 2A), as observed in monkeys [1]. On the other hand, OD columns [9] and CO blobs in cats [10] are observed after eye opening. This suggests that OD columns and CO blobs in cats are formed under positive correlation in between-eye activities, because retinal neurons of the right and left eyes focusing on the same visual point should have positive correlation in the activities because of their same pattern vision. When h takes a

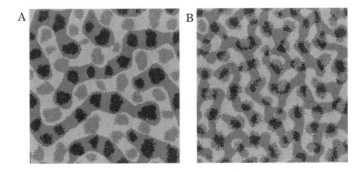

FIG. 2A,B. Patterns of synaptic terminals. **A** Patterns of afferent inputs after self-organization. Parameters used here are $h = 0.0$, $g = 0.4$, $\gamma_{L,A} = \gamma_{R,A} = 2.5$, $\gamma_{L,B} = \gamma_{R,B} = 0.4$ and $c_A = 0.006$, $c_B = 0.0005$. **B** Patterns of synaptic inputs when the correlation in between-eye activities h takes a positive value ($h = 0.4$)

positive value ($h = 0.4$), nearby blobs tend to merge across the OD borders (Fig. 2B), like the CO staining patterns in cats [2].

Discussion

In diurnal primates, koniocellular (K) and magnocellular (M) LGN cells send their outputs inside CO blobs, and parvocellular (P) LGN cells send their outputs both inside and outside CO blobs [11]. In diurnal primates, therefore, group A neurons in the present model consist of K, M, and P cells, and group B neurons consist of P cells. In nocturnal primates, on the other hand, K and P cells send their outputs inside CO blobs, and M cells send their outputs both inside and outside CO blobs [11]. Thus, in nocturnal primates, group A neurons in the model consist of K, M, and P cells, and group B neurons consist of M cells. In cats, W and Y cells send their outputs inside CO blobs, and Y and X cells send their outputs outside the blobs. In cats, therefore, group A neurons in the model consist of W and Y cells, and group B neurons consist of X and Y cells.

References

1. Horton JC, Hubel DH (1981) Regular patchy distribution of cytochrome oxidase staining in primary visual cortex of macaque monkey. Nature 292:762–764
2. Murphy KM, Jones DG, van Sluyters RC (1995) Cytochrome-oxidase blobs in cat primary visual cortex. J Neurosci 15:4196–4208
3. Horton JC, Hocking DR (1996) Anatomical demonstration of ocular dominance columns in striate cortex of the squirrel monkey. J Neurosci 16:5510–5522
4. Tanaka S (1990) Theory of self-organization of cortical maps: mathematical framework. Neural Netw 3:625–640
5. Nakagama H, Saito T, Tanaka S (2000) Effect of imbalance in activities between ON- and Off-center LGN cells on orientation map formation. Biol Cybern 83:85–92
6. Wong-Riley MTT (1994) Primate visual cortex. Dynamic metabolic organization and plasticity revealed by cytochrome oxidase. In: Rockland PA, Rockland KS (eds) Cerebral cortex vol. 10. Plenum Press, New York, pp 141–200
7. Horton JC, Hocking DR (1996a) An adult-like pattern of ocular dominance column in striate cortex of newborn monkeys prior to visual experience. J Neurosci 16:1791–1807
8. Purves D, Lamantia A (1993) Development of blobs in the visual cortex of macaques. J Comp Neurol 334:169–175
9. Rathjen S, Löwel S (2000) Early postnatal development of functional ocular dominance columns in cat primary visual cortex. Neuroreport 11(11):2363–2367
10. Murphy KM, Duffy KR, Jones DG, et al. (2001) Development of cytochrome oxidase blobs in visual cortex of normal and visually deprived cats. Cereb Cortex 11:1122–1135
11. Casagrande VA, Kaas JH (1994) The afferent, intrinsic and efferent connections of primary visual cortex in primates. In: Rockland PA, Rockland KS (eds) Cerebral cortex vol. 10. Plenum Press, New York, pp 201–259

Long-Range Horizontal Connections Assist the Formation of Robust Orientation Maps

Nicolangelo Iannella, Jerome Ribot, and Shigeru Tanaka

Key words. Long-range horizontal connections, Spiking neurons, Visual cortex

Introduction

In layer II/III of cat area 17, pyramidal cells and large basket cells are known to form long-range excitatory and inhibitory connections, respectively. The long-range horizontal connections (LRHCs) of pyramidal cells have the tendency to contact other cells with similar orientation preference, in a clustered or anisotropic fashion, up to 8 mm away from the cell body [1]. They are believed to be modulatory in nature and permit cortical neurons to integrate information from outside their classical receptive field. Furthermore, electrophysiological studies have illustrated that contextual stimuli mediate both facilitative and suppressive effects, thereby suggesting a functional role [1, 2].

A combined anatomical, electrophysiological, and optical imaging experiment [3] has quantified that there may be a geometric relationship between patchy long-range connections of pyramidal cells and the domains of the orientation map. Furthermore, this study illustrated that axons of large basket cells display no specificity in functional preference nor do they extend as far as those of pyramidal cells do.

Recently, theoretical studies have started to investigate what effects LRHC have on the behavior of model networks composed of spiking neurons. One theoretical study has demonstrated that LRHC connections can mediate similar contextual effects, as seen in experiment [4]. From the theoretical point of view, apart from contextual modulation, could LRHC play some other role? In this brief paper, we will demonstrate that incorporating LRHC into large-scale networks of spiking neurons can promote a robust orientation map.

Laboratory for Visual Neurocomputing Brain Science Institute, RIKEN, 2-1 Hirosawa, Wako, Saitama 351-0198, Japan

Methods and Materials

The Model

We constructed a large-scale model network of cat area 17, composed of spiking excitatory and inhibitory neurons and geniculo-cortical afferent inputs, predetermined via self-organization. Both excitatory and inhibitory cortical neurons were modeled as integrate-and-fire units, using experimentally reported membrane time constants. The network contains more than 4000 neurons and over 3 million connections. Periodic boundary conditions were assumed. For simplicity, the model consists of a single cortical layer of 48×48 pixels, representing a 2.4-mm by 2.4-mm patch of area 17. For computational tractability, a single pair of excitatory and inhibitory neurons approximates the localized population of cells underneath a single pixel (of area 2500 μm^2). Only α-amino-3-hydroxy-5-methyl-4-isoxazole-mediated excitation and γ-aminobutyric acid-A mediated inhibition were included. Drifting sinusoidal grating patterns, moving in eight directions, were used as visual stimuli.

Lateral Geniculate Nucleus (LGN) Model of Afferent Firing Rates

Four types of geniculate cells, ON-center nonlagged, ON-center lagged, and their OFF-center counterparts, were included to provide afferent inputs to the cortical layer. The model LGN consists of four distinct layers composed of a single cell type, where each layer contains 576 cells arranged in a 24×24 lattice. The afferent spike sequences that drives our cortical neurons were modeled as Poisson processes whose firing rates are calculated using the convolution between the geniculate cell's spatiotemporal receptive field, and the visual stimulus. If the firing rate was negative, then no spike was fired.

Connectivity

Both geniculo-cortical and inhibitory connection weights were modeled as normalized Gaussians, while only excitation formed both short-range and long-range connections. The long-range connections used here were an idealization of those seen from anatomical studies. Like the geniculo-cortical and inhibitory connections, short-range excitatory connections were modeled as normalized Gaussians. To implement long-range circuitry, we imposed a few assumptions. First, observing the resultant orientation map in the absence of cortical connections, one can deduce that the approximate size of a single hyper-column is a region of 16×16 pixels, and, furthermore, the underlying coordinate system is approximately Cartesian. Second, the spatial extent of each patch is identical to the short-range excitatory connections. Using these assumptions, a 2-dimensional template incorporating both short- and long-range interactions, whose underlying Cartesian coordinate system was spatially matched, pixel-by-pixel, to that of the model cortex, was adopted. Short-range connections were centered at the origin of the template, while patches were arranged on a regular lattice, where the distance between nearest neighbor patch centers in either ordinate direction was 16 pixels. LRHCs were selected by centering the origin of the template to cortical pixel j, and comparing the orientation preference of cortical pixel j to those in

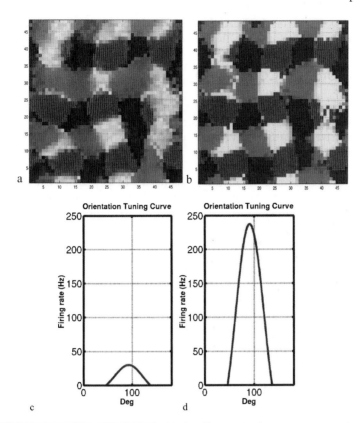

FIG. 1a–d. Increasing the patch efficacy had little effect on the domain structure of the orientation map and corresponding orientation tuning curves. **a, b** Patchy long-range connections are present and their efficacy is systematically increased. **c, d** Orientation tuning curves of a cell taken from the center of each map, respectively

the patch. If the absolute difference between orientation preferences was ≤30°, then a connection was present. Furthermore, LRHC contacted both excitatory and inhibitory units.

Results

In order to demonstrate that LRHCs promote "robust" orientation maps, one simply needed to demonstrate that the basic structure of the orientation domains effectively remains fixed when the synaptic efficacies of the patches are increased. Figure 1a,b demonstrates that increasing the efficacies of connections originating from the patches has little effect on the overall structure of the orientation map. We will refer to this as the patch efficacy from now on. Figure 1a depicts the orientation map when the patch efficacy is set to one-eightieth of the maximum strength of short-range excitation. In Fig. 1b, the patch efficacy has been increased by 100% and the orientation map seems to have "crystallized," but there is little change in its global structure.

Discussion and Conclusion

The role of long-range horizontal connections within the cortex is not fully understood. Previous studies have demonstrated that they could mediate both facilitative and suppressive contextual interactions, as well as contrast saturation and contrast-dependent length tuning. Here, we have explored another possibility, the effect they have on maintaining global features of the orientation map in large-scale networks, based on spiking neurons. We have demonstrated that they could effectively lock or clamp the global structure of the orientation map. At first glance, one may think that having an orientation map, whose structure effectively remains invariant when the patch efficacy varies, may not be an interesting phenomenon. Now, we will add an extra twist to the story. Inspecting the orientation-tuning curve of several cells about the center of each map depicted in Fig. 1a,b, respectively, revealed that the maximum firing rate at the preferred orientation has been dramatically increased. Additionally, we noted that their respective tuning curve had similar profiles, suggesting that patchy connections seem to implement a particular form of boosting at the preferred orientation. This result may have some important ramifications from an engineering perspective. First, not only could they mediate both facilitative and suppressive modulation but also the same style of circuitry may allow a neuron to operate under a wide operational range. Furthermore, it seems to suggest that patchy connections could implement a form of "functional invariance," in which the behavior of the processing unit could remain robust under plastic changes in patch efficacy. In conclusion, the result of this brief paper and that by others indicates that such LRHCs may play multiple roles in the processing of information in the primary visual cortex.

Acknowledgments. The authors would like to thank Mr. Masanobu Miyashita from FRL, NEC Corporation for providing us with afferent input data.

References

1. Gilbert CD, Wiesel TN (1989) Columnar specificity of intrinsic horizontal and cortico-cortical connections in cat visual cortex. J Neurosci 9:2432–2442
2. Hirsch JA, Gilbert CD (1991) Synaptic physiology of horizontal connections in the cat's visual cortex. J Neurosci 11:1800–1809
3. Kisvárday ZF, Tóth E, Rausch M, et al. (1997) Orientation-specific relationship between populations of excitatory and inhibitory lateral connections in the visual cortex of the cat. Cereb Cortex 7:605–618
4. Somers DC, Todorov EV, Siapas AG, et al. (1998) A local circuit approach to understanding integration of long-range inputs in primary visual cortex. Cereb Cortex 8:204–217

Experience-Dependent Self-Organization of Visual Cortical Receptive Fields and Maps

Masanobu Miyashita[1,2] and Shigeru Tanaka[2]

Key words. Self-organization, Visual cortex, Cortical column, Spatiotemporal receptive field, Visual experience

Introduction

It has been revealed that neurons that selectively respond to similar stimulus features are systematically arranged in the visual cortex to form columnar structures such as ocular dominance, orientation, and direction-of-motion columns [1, 2]. Since Blakemore and Cooper [3] reported that preferred orientations were strongly biased to an experienced orientation for kittens that had stayed inside a cylindrical room and been exposed to a stripe painted on the inner wall of the room, many investigators [3–10] have attempted to elucidate the role of visual experience in the emergence of orientation- and direction-selective cells and related columnar structures. On the other hand, Crair et al. [10] have suggested that visually driven activity is not required for orientation map formation and full maturation of orientation selectivity. It seems to be accepted that the basic structure of orientation maps is unexpectedly robust against restricted visual experience [5, 6, 10].

In this study, we investigated theoretically the influence of visual experience on the formation of cortical receptive fields and columnar structures, using our mathematical model for the activity-dependent self-organization of geniculo-cortical afferent inputs based on the Hebbian coincidence of pre- and postsynaptic activities [7, 8]. We assumed that cortical cells receive afferent inputs from four types of lateral geniculate nucleus (LGN) cells: ON- and OFF-center lagged cells, and ON- and OFF-center nonlagged cells [8, 11]. To determine the possible effects of visual experience, we carried out computer simulations for different ratios of contributions between instruction-based learning and correlation-based learning, both of which are derived from the Hebbian coincidence. Simulated results suggest that visual experience is not necessarily required for the formation of cortical receptive fields and columnar

[1] Fundamental Research Laboratories, NEC, 34 Miyukigaoka, Tsukuba 305-8501, Japan
[2] Laboratory for Visual Neurocomputing, Brain Science Institute, RIKEN, 2-1 Hirosawa, Wako, Saitama 351-0198, Japan

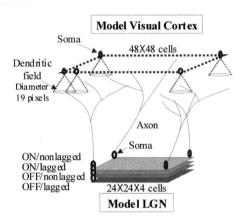

Model Visual Cortex

Soma

48X48 cells

Dendritic field

Diameter 19 pixels

Axon

Soma

ON/nonlagged
ON/lagged
OFF/nonlagged
OFF/lagged

24X24X4 cells

Model LGN

FIG. 1. Schematic diagram of model lateral geniculate nucleus (*LGN*) and cortex

structures, but it facilitates their maturation. Furthermore, it was shown that during the presentation of only a vertically oriented moving grating, the iso-orientation domains of the vertical orientation occupied more cortical territory than the other orientations. However, this overrepresentation effect was suppressed as the contribution of correlation-based learning increased. Thus, it is implied that the instruction-based learning and correlation-based learning work together for orientation map formation.

Method

In the present model, the pattern of geniculo-cortical connections is given by maximizing the Hebbian coincidence between postsynaptic membrane depolarization and presynaptic cell firing under certain constraints on the number of synaptic connections from individual presynaptic cells [7, 8]. We took into account retinotopic projection from retina to LGN and four types of LGN cells: lagged and nonlagged types for ON- and OFF-center cells (Fig. 1).

The Hebbian coincidence can be generally given as the sum of two terms: The first term indicates the product of average activities at LGN and the visual cortex driven by visual input, and the second term indicates activity correlation caused by fluctuation in firing evoked by spontaneous retinal burst activities [12] and/or cortico-geniculate feedback signals. It can be said that the first term is related to instruction-based learning, and the second term is related to correlation-based learning. Note that instruction-based learning is of the order of the square of the average firing rate of LGN cells, whereas correlation-based learning is proportional to the average firing rate when we assume LGN cells' firing to obey the Poisson process. This indicates that instruction-based learning is dominant for higher firing rate and correlation-based learning is dominant for lower firing rate. In simulations, eight grating patterns were presented to the model visual field, which moved in the directions from 0° to 315°. The periodic boundary conditions were imposed on the model LGN and cortex.

Results

Figure 2a1,a2 illustrates typical orientation and direction maps obtained from simulation with the correlation-based learning only. The segregation of orientation and direction preferences into patches in these pictures indicates that cortical maps can form without pattern vision, but iso-orientation and iso-direction domains are not clearly delineated. These immature cortical maps have been observed in the visual cortex of cats and ferrets immediately after eye opening or during binocular lid suture [10, 13]. We also found spatiotemporal inseparable and separable simple-cell-like receptive fields self-organized in the model cortex, although orientation and direction tunings were broader than those experimentally observed in adult cats.

When we continued to conduct the earlier-mentioned simulation with the instruction-based learning only (Fig. 2b1,b2), the model cortical domains were further segregated into clear iso-orientation and iso-direction domains. Interestingly, the basic structures in the arrangement of orientation and direction preferences in Figs. 1 and 2 are similar. This suggests that general structural features of cortical maps are not changed by switching conditions from binocular deprivation to normal visual experience.

Next, we carried out simulations during the presentation of only vertical orientation. When we assumed only correlation-based learning, most of the model cortical cells responded rigorously to the vertical orientation, and the experienced orientation occupied a much larger cortical territory than the others (Fig. 2c). However, this over-representation effect became more strongly suppressed as the contribution of correlation-based learning increased (Fig. 2d–f).

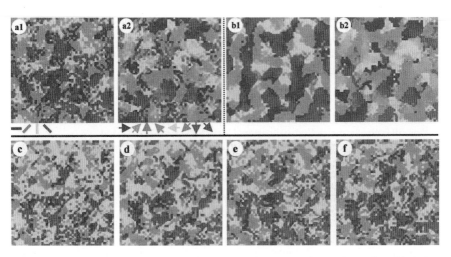

Fig. 2. **a1,a2** Simulated orientation and direction angle maps with correlation-based learning only. **b1,b2** Simulated orientation and direction angle maps with instruction-based learning only. **c–f** Simulated orientation angle maps when only vertically oriented gratings were presented with different ratios (r) in the magnitude of correlation-based learning to instruction-based learning: $r = 0$ for **c**; $r = 0.5$ for **d**; $r = 1$ for **e**; and $r = 2$ for **f**

Discussion and Conclusion

We found in our simulated results that receptive fields elongated along the axis of their preferred orientation (not shown), and hence their tuning width became narrower. Crair et al. [10] have emphasized that the formation of cortical maps is innate, but the role of visual experiences is essential for maintaining the responsiveness and selectivity of cortical cells. Our model suggests that visual experience not only maintains but also facilitates the maturation of receptive fields and cortical maps that have been formed under binocular deprivation. This theoretical consequence is consistent with White et al.'s experimental result [13].

When kittens were reared in darkness except for several hours a day during which they were exposed to a stripe pattern, only slight overrepresentation was observed [5, 6]. In such studies, even if the contribution of instruction-based learning is larger than that of correlation-based learning during the exposure to a single orientation, correlation-based learning is suspected to have a substantial influence on map formation during the period of dark rearing. If this is the case, the overrepresentation should be suppressed. In Tanaka et al.'s very recent experiment, goggles-mounted kittens that could experience only vertical orientation whenever they were awake exhibited extremely large iso-orientation domains for the vertical orientation (see the chapter by S. Tanaka, J. Ribot, and K. Imamura, this volume). For those kittens, it is unlikely that correlation-based learning had a strong influence on the map formation since they were never placed in darkness during development. Taken together, the results show that the degree of overrepresentation of the experienced orientation is regulated by the ratio of the contribution of instruction-based learning to that of correlation-based learning to orientation map development.

Acknowledgments. This study was performed in part through Special Coordination Funds for Promoting Science and Technology from the Ministry of Education, Culture, Sports, Science and Technology, Japan.

References

1. Hubel DN, Wiesel TN (1962) Receptive fields, binocular interaction and functional architecture in the cat's visual cortex. J Neuro Physiol 160:106–154
2. Hubener M, Shoham D, Grinvald A, et al. (1997) Spatial relationships among three columnar systems in cat area 17. J Neurosci 17:9270–9284
3. Blakemore C, Cooper GF (1970) Development of the brain depends on visual environment. Nature 228:477–478
4. Rauschecker JP, Singer W (1981) The effects of early visual experience on the cat's visual cortex and their possible explanation by Hebb synapses. J Physiol 310:215–239
5. Stryker MP, Sherk H (1975) Modification of cortical selectivity in the cat by restricted visual experience: a reexamination. Science 190:904–906
6. Sengpiel F, Stawinski P, Bonhoeffer T (1999) Influence of experience on orientation maps in cat visual cortex. Nat Neurosci 8:727–732
7. Miyashita M, Tanaka S (1992) A mathematical model for the self-organization of orientation columns in visual cortex. Neuroreport 3:69–72
8. Miyashita M, Kim DS, Tanaka S (1997) Cortical directional selectivity without directional experience. Neuroreport 8:1187–1191

9. Miller KD (1994) A model for the development of simple cell receptive fields and the ordered arrangement of orientation columns through activity-dependent competition between ON- and OFF-center inputs. J Neurosci 1: 409–441
10. Crair MC, Gillespie DC, Stryker MP (1998) The roles of visual experience in the development of columns in cat visual cortex. Science 279:566–570
11. DeAngelis GC, Ohzawa I, Freeman RD (1995) Receptive-field dynamics in the central visual pathways. Trends Neurosci 18:451–458
12. Meister M, Wong ROL, Baylor DA, et al. (1991) Synchronous bursts of action potentials in ganglion cells of the developing mammalian retina. Science 252:939–943
13. White LE, Coppola DM, Fitzpatrick D (2001) The contribution of sensory experience to the maturation of orientation selectivity in ferret visual ·cortex. Nature 411: 1049–1052

Distributions of Binocular Phase Disparities in Natural Stereopair Images

Sakiko Nouka and Izumi Ohzawa

Key words. Binocular disparity, Natural image, Stereopsis

Stereopsis is an ability to gauge depth from a pair of two-dimensional retinal images. Because the eyes are separated laterally, the two retinal images differ slightly from each other, with images of objects shifted mainly along the horizontal dimension. This shift is known as binocular disparity. Previous studies suggest that binocular disparity information may be encoded by simple cells with varying degrees of dissimilarity between left and right receptive field (RF) profiles [1]. Furthermore, there is an anisotropy in the distribution of RF dissimilarities (as quantified by RF phase difference) with respect to the neuron's preferred orientation. Neurons that prefer near vertical orientation (which encode horizontal disparities) could have similar or dissimilar RF profiles. However, those that prefer near horizontal orientation (which encode vertical disparities) have highly similar RF profiles for the two eyes. Therefore, the visual system appears to employ a highly efficient encoding scheme tailored for statistical properties of the binocular visual environment.

In this study, we have analyzed the degree of orientational anisotropies actually present in natural binocular environment in the hope of taking advantage of such anisotropies for developing an efficient system of transmission and storage of stereo images. For this purpose, we measured the distribution of binocular phase disparities for natural stereopair images at a variety of orientations and spatial frequencies.

Our results show that there is no substantial orientational anisotropy in the distribution of binocular phase disparities in natural stereo images. The average amount of vertical disparity information can reach 80% of that for horizontal disparities. These results indicate that there is little hope of gaining efficiency for stereo image encoding by taking advantage of orientational anisotropy in images. Therefore, it is somewhat puzzling as to how the visual system could develop orientational bias in the phase disparity distributions.

Reference

1. DeAngelis GC, Ohzawa I, Freeman RD (1991) Depth is encoded in the visual cortex by a specialized receptive field structure. Nature 352:156–159

Graduate School of Engineering Science, Department of Biophysical Engineering, and Graduate School of Frontier Biosciences, Osaka University, Japan

A Network Model for Figure/ Ground Determination Based on Contrast Information

Haruka Nishimura[1] and Ko Sakai[2]

Key words. Figure/ground determination, Model, Contrast, V2

Introduction

Discrimination of a figure from its ground is essential for the perception of surface, which is fundamental for the recognition of shape, spatial structure, and objects. Recent physiological studies have reported that neurons in monkey's V2 and V4 show activities corresponding to the direction of figure [1–3]. Zhou et al. [1] have distinguished types of neurons that showed selectivity to border ownership and/or contrast, about 80% of which were sensitive to contrast polarity. Figure 1a illustrates the distribution of the neuronal selectivity in V1, V2, and V4. The selectivity to border ownership and contrast was observed in stimuli with single squares, C-shaped figures, and two overlapping squares, as shown in Fig. 1b. However, underlying neural mechanisms for figure determination have not been clarified.

We propose a network model for determining figure/ground based on the combination of contrast information. Although a number of models that utilized T junctions and ownership junctions have been proposed (e.g., [4]), neurons selective to such junctions have not been reported. The strong contrast dependency apparent in the figure-selective neurons in V2 and V4 led us to expect that specific combinations of local contrast could be a basis for the mechanism that determines the direction of figure. We carried out the simulations of the model on the Nexus neural simulator [5] in order to examine whether in fact the model reproduces the responses of V2 and V4 neurons reported by Zhou et al. [1]. The stimuli for the simulations included C-shaped figures and two overlapping squares. The results of the simulations show that the model is capable of determining figure/ground solely from the combinations of local contrast that can be gathered through horizontal connections apparent in early vision.

[1] Graduate School of Systems and Information Engineering, and [2] Institute of Information Sciences and Electronics, University of Tsukuba, Tennodai, Tsukuba 305-8573, Japan

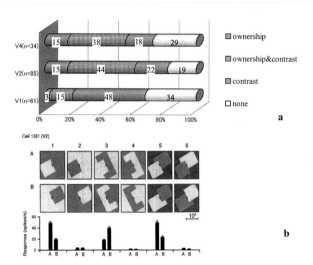

FIG. 1a, b. **a** The distribution of neuronal selectivity in V1, V2, and V4. There are four types of neurons selective to (1) border ownership, (2) border ownership and polarity of edge contrast, (3) polarity of edge contrast, and (4) others (replotted from data in Zhou et al. [1]). The majority of neurons are sensitive to contrast polarity. **b** Response of a cell recorded in V2, which is selective to both border ownership "left," and the polarity of edge contrast "bright-dark." The selectivity of the cell was retained in several stimuli including single squares, c-shaped figures, and two overlapping squares. (Reproduced from Zhou et al. [1] with permission. Copyright 2000 by the Society of Neuroscience)

Model

We propose a network model for figure/ground determination based on the combination of contrast information. The model consists of three major stages. The model first extracts local contrast and its orientation from stimuli by detecting intensity changes with V1-simple-cell-like units. The next stage realizes the selectivity to border ownership that depends on contrast polarity. Finally, the third stage shows the border ownership regardless of contrast polarity. The following sections describe major functions of each stage, and Fig. 2a illustrates the architecture of the model.

Detection of Edge Contrast

The units in this stage (Fig. 2a B) realize the selectivity to edge contrast by taking the convolution with Gabor filters. For detecting "light-dark" edges, we use $G_1(x,y)$ that is responsive if intensity decreases from left to right along horizontal orientation (0° phase). For detecting "dark-light" edges, we use $G_2(x,y)$ that is antiphase (180°) of $G_1(x,y)$. The response of the units is given by

$$O_1(x_0, y_0) = \sum G_1(x, y) * input(x_0, y_0)$$
$$O_2(x_0, y_0) = \sum G_2(x, y) * input(x_0, y_0)$$

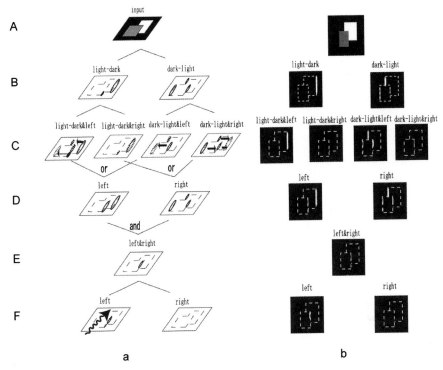

FIG. 2a, b. The architecture and simulation results. **a** The architecture of the model that determines figure/ground. *A*, An input stimulus, an *overlapping dark square in front of a light square*. *B*, Units detecting edge contrast with Gabor filters, "light-dark" and "dark-light." *C*, Units coding both polarity of edge contrast and border ownership. The detailed mechanism is shown in Fig. 3. A "light-dark & left" unit detects the part that edge contrast is "light-dark" and that ownership is "left." It detects contrast on the left as indicated by *arrows*. *D*, Units realizing the selectivity to border ownership invariant to contrast. A "left" unit adds the responses of "light-dark & left" and "dark-light & left" units. A "right" unit adds the responses of "light-dark & right" and "dark-light & right" units. Determining figure/ground is completed in this stage if an input is a single figure. *E*, The *overlapped part* shows ownership to both left and right and is thus ambiguous. *F*, The ownership determined at D propagates to the overlapped part. **b** The response of each stage. *Brighter grey* shows stronger activation. *Dashed lines* are the contours of the input stimulus. *Occluding contour* shows the response of "left" at the last

where *input* (x_0, y_0) is a stimulus, and * shows convolution. $O_1(x,y)$ and $O_2(x,y)$ represent the response to "light-dark" edges and "dark-light" edges, respectively.

Selectivity to Border Ownership with Specific Contrast Polarity

The units in this stage (Fig. 2a *C*) realize the selectivity to edge contrast and border ownership. There are four kinds of selectivity: "light-dark & left," "light-dark & right," "dark-light & left," and "dark-light & right." If contrast is detected in the direction of search, the border ownership is considered to exist in that direction. Figure 3 illustrates the detailed mechanism.

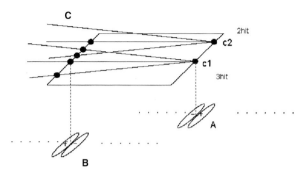

Fɪɢ. 3. An illustration of the mechanism that establishes the selectivity to edge contrast "dark-light" and border ownership "left." A "dark-light" filter, A, detects a "dark-light" edge, and a "light-dark" filter, B, detects a "light-dark" edge. The mask C realizes a search for contrast change to the left, which models lateral connections in early vision. A unit, $c1$, takes inputs from several "light-dark" units on the left (B) through a mask C, together with inputs from "dark-light" units (A). Therefore, $c1$ responds to "dark-light" and "left"

Selectivity to Border Ownership Invariant to Contrast Polarity

This stage realizes the selectivity to border ownership without specific contrast polarity by taking logical OR between the units with the same ownership but different contrast polarity at the same retinotopic position as in the previous stage (Fig. 2a D). If a stimulus consists of a single figure without overlap, this stage makes correct judgment of the ownership. However, this is not the case if a stimulus is two overlapping squares, for which the next stage is necessary.

Border Ownership of Overlapping Figures

Zhou et al. [1] have reported that neurons in V2 and V4 correctly signal the figure direction in two overlapping squares (Fig. 1b). The units in the previous stage show both right and left ownership at an occluding contour because the change in contrast is detected both right and left of the occluding contour. Therefore, those units are confused, and incapable of determining correct ownership (Fig. 2a E). In this stage (Fig. 2a F), the ownership of the occluding contour is determined from the signals propagated from the extension of the contour that goes beyond the occluded figure so that no overlap exists and the ownership is determined without ambiguity. The ownership of the contour on the extension propagates to the occluding part through collinear connections. The collinear connection is suggested to play a key role in the perception of contour and orientation in the primary visual cortex [6, 7]. We assume the similar connections in V2.

Results

We carried out the simulations of the model on the Nexus neural simulator [5]. An example of the results is shown in Fig. 2b in which the ownership of the overlapped part of a stimulus was determined successfully as "left" (Fig. 2b F). The model reproduces major properties of the V2 and V4 neurons reported by Zhou et al. [1], including the figure-direction selectivity in C-shaped figures and two overlapping squares. The results indicate that the model is capable of determining figure/ground solely from the combinations of local contrast information gathered through colinear and lateral connections.

Discussion

The present study has shown that local contrast information is capable of determining figure direction. We investigate further whether other cues are involved in the ownership detection. Zhou et al. [1] have reported that some neurons in V2 and V4 selective to figure direction responded correctly to figure even if only a partial view of a single square was presented so that the other side of square was invisible. They have also reported that the neurons responded correctly to figure if a stimulus had an elliptic shape. This result suggests that something other than local contrast or junctions might play a role in the figure-selective neurons. Our preliminary results show that collinear connections in model V2, which has similar connectivity to horizontal connections apparent in V1, propagate the signals of figure direction so that a partial view of an object, including those with an arbitrary shape, could give activities similar to the neuronal responses reported.

References

1. Zhou H, Friedman HS, von der Heydt R (2000) Coding of border ownership in monkey visual cortex. J Neurosci 20:6594–6611
2. Von der Heydt R, Zhou H, Friedman HS (2000) Representation of stereoscopic edges in monkey visual cortex. Vision Res 40:1955–1967
3. Bakin JS, Nakayama K, Gilbert CD (2000) Visual responses in monkey areas V1 and V2 to three-dimensional surface configurations. J Neurosci 20:8188–8198
4. Finkel LH, Sajda P (1994) Constructing visual perception. Am Sci 82:224–237
5. Sajda P, Sakai K, Yen SC, et al. (1994) NEXUS: A neural simulator for integrating top-down and bottom-up modelling. In: Skrzypek J (ed) Neural network simulation environments. Kluwer, Boston, pp 29–45
6. Yen SC, Finkel LH (1998) Extraction of perceptually salient contours by striates cortical networks. Vision Res 38:719–774
7. Sakai K, Hirai Y (2002) Neural grouping and geometric effect in the determination of apparent orientation. J Opt Soc Am A 19:1049–1062

Neural Responses to the Uniform Surface Stimuli in the Visual Cortex of the Cat

Toshiki Tani[1,2], Isao Yokoi[1], Minami Ito[1], Shigeru Tanaka[3], and Hidehiko Komatsu[1,2]

Key words. Uniform surface, Visual cortex of the cat, Optical imaging

In the early stages of the visual cortex (areas 17 and 18), it is well known that neurons respond to bar and contour stimuli. However, several recent studies reported that some neurons in this stage respond to uniform plane stimuli that cover the classical receptive fields [1, 2]. In order to better understand the neuronal mechanisms underlying surface representation in the early visual stages, we explored the distribution of activation to the uniform plane stimuli by using optical imaging of intrinsic signals and extracellular recordings from areas 17 and 18 of anesthetized cats. We also intended to examine the distribution of such activities in relation to the orientation preference map in the same areas.

The optical imaging studies revealed the presence of a series of spots in area 18 with strong activation by uniform plane stimuli. These spots were located parallel to the border between areas 17 and 18. They overlapped with the singular points of the orientation preference map. Extracellular recordings from superficial layers showed that more than 80% of neurons recorded within the spots were responsive to the uniform plane stimulus. These neurons also responded to low spatial frequency gratings. Outside of the spots, smaller proportion of neurons (<40%) responded to the uniform plane stimuli and these neurons tended to be located near the singular points.

We suggest that neurons responding to the uniform plane stimuli form clusters in area 18 and that these neurons may be involved in an integration of surface information and contour information. The specific localization of these clusters suggests that they play some role in the integration of a large surface across the vertical meridian.

[1] Laboratory of Sensory and Cognitive Information, National Institute for Physiological Sciences, Myodaiji, Okazaki, Aichi 444-8585, Japan
[2] Department of Physiological Sciences, The Graduate University for Advanced Studies, Okazaki, Aichi 444-8585, Japan
[3] Laboratory for Visual Neurocomputing, Brain Science Institute, RIKEN, 2-1 Hirosawa, Wako, Saitama, Japan

References

1. Rossi AF, Rittenhouse CD, Paradiso MA (1996) The representation of brightness in primary visual cortex. Science 273:1104–1107
2. Kinoshita M, Komatsu H (2001) Neural representation of the luminance and brightness of a uniform surface in the macaque primary visual cortex. J Neurophysiol 86:2559–2570

Cascaded Energy Model as the Basis for Neural Processing in V2

Takeshi Ikuyama and Izumi Ohzawa

Key words. Texture, Generalized receptive field, Second-order response

What happens to visual information beyond the simple and complex cells of V1? Qualitatively, we know that two major cortical streams, namely, ventral and dorsal streams, originate from V1. Little is known, however, about specific neural wirings that produce neurons in areas beyond V1. In this study, we have modeled possible neural computations by combining the output of cells in V1, and have compared predictions of the models to existing data from V2 neurons.

The model we propose has a hierarchical organization that repeats, in a cascaded manner, the simple-to-complex hierarchy in V1 as implemented by an energy model [1, 2]. It is therefore called a "cascaded energy model." The first stage in V2 is modeled by combining the output of V1 complex cells linearly with a spatial distribution of weights according to a two-dimensional Gabor function. Neurons in this stage possess properties of simple cells except that they respond to texture-defined edges instead of luminance-defined borders. Therefore, it is appropriate to name these V2 neurons as "texture simple cells." The stage following texture simple cells is a layer of "texture complex cells" and each of these cells combines the output of four texture simple cells according to the energy model organization.

Examinations of responses of texture complex cells show that these neurons respond well to texture edges, or more generally, spatial modulations of the contrast of texture. Predicted responses agree remarkably well with existing data from V2 neurons. These model neurons also have properties of end stopping or surround suppression that is common in V2 neurons. In addition, they respond to stimulus edges found in a variety of illusory or subjective contour stimuli. In summary, the cascaded energy model we propose appears to provide a unified description of many properties of V2 neurons previously thought to be unrelated.

Department of Biophysical Engineering, Graduate School of Engineering Science and Graduate School of Frontier Biosciences, Osaka University, 1-3 Machikaneyama, Toyonaka, Osaka 560-8531, Japan

238

References

1. Adelson EH. Bergen JR (1985) Spatiotemporal energy models for the perception of motion. J Opt Soc Am A 2:284–299
2. Ohzawa I, DeAngelis GC, Freeman RD (1990) Stereoscopic depth discrimination in the visual cortex: neurons ideally suited as disparity detectors. Science 249:1037–1041

Behavioral and Physiological Study of Color Processing in Human Brain

Hitoshi Sasaki[1], Sumie Matsuura[2], Akiko Morimoto[1], and Yutaka Fukuda[1]

Key words. Simple reaction time, Discrimination time, Alpha-blocking, Color detection, Cerebral hemispheric dominance

Introduction

Although color processing is one of the important features in visual information processing in humans, little is known about color processing in the human brain. In the present study, we asked two questions using human subjects. First, we asked how color stimulation of different hues (wavelengths) is processed within the brain. Second, we asked whether or not hemispheric differences exist between the right and left hemispheres in the processing of color stimulation. To answer these questions, we presented chromatic stimuli, matched photometrically in luminance, on an achromatic background or on an achromatic circle of the same size with the same luminance, and measured simple and discriminative reaction times (RTs) as well as latencies of alpha-blockings in EEG induced by these color stimuli.

Effects of Hue on Simple Reaction Time

RT is the minimal time required for detection of sensory input response and selection, stimulus identification for motor output. It has been widely used to evaluate brain functions including visual information processing [1]. Although many studies have been reported as to the effect of hue on simple RT, a consistent result has not been reported for the differences among different hue stimuli [2–4]. In the present experiments, the subjects were seven male undergraduate students (aged 20–23 years) with normal or corrected normal vision. They were instructed to press a button as fast as possible after they perceived the visual target. All visual stimuli were presented on a cathode ray tube (CRT) display with a uniform achromatic background of 24 cd/m^2 ($10° \times 10°$). As a fixation point, an achromatic 2° circle of 10 cd/m^2 was always

[1] Department of Physiology and Biosignaling, Osaka University Graduate School of Medicine, 2-2 Yamadaoka, Suita, Osaka 565-0871, Japan
[2] San Cuore, 4-22-2 Nakayamate St., Chuo-ku, Kobe, Japan

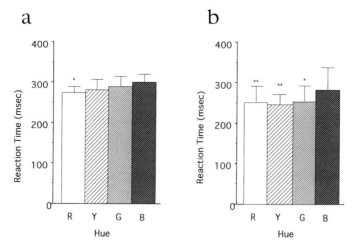

FIG. 1a,b. Effects of hue on simple reaction time (RT). **a** Simple RT to chromatic stimuli (red (R), 635 nm; yellow (Y), 580 nm; green (G), 548 nm; blue (B), 463 nm) in seven subjects. There was a significant difference between RTs to red and blue ($t(6) = 2.885, P < .05$). Mean with SD. **b** Simple RT to chromatic stimuli with different main wavelengths (R, 600 nm; Y, 577 nm; G, 537 nm; B, 465 nm) in ten different subjects. RT to blue was significantly longer than any other hue (*$P < .05$; **$P < .01$). Mean with SD

presented at the center of the display. Target stimuli were 2° circles with four different main wavelengths [(red (R), 635 nm; yellow (Y), 580 nm; green (G), 548 nm; blue (B), 463 nm)]. Luminance of these stimuli was matched to the achromatic circle of 10 cd/m^2 by flicker photometry. Saturation was adjusted to a blue stimulus of 60%. With a random delay of 3–5 s, following a ready signal (tone 55 dB, with a background level of 42 dB, SPL), one of the target stimuli was presented on the fixation point with an intertrial interval (ITI) of 15 s (10–20 s). As shown in Fig. 1a, simple RT to blue was significantly longer than that to red. Similar results were obtained in another experiment using different main wavelengths (R, 600 nm; Y, 577 nm; G, 537 nm; B, 465 nm; Fig. 1b). Here again, the RT to blue was significantly longer than that to red, yellow, or green.

Effects of Hue on EEG Alpha-Blocking

Alpha waves in the EEG appear dominantly in the resting state, and reduce their amplitude in response to visual stimulation (Fig. 2a). This phenomenon is called alpha-blocking, and can be used as a sign of changes in arousal level of the brain. Ten undergraduate students (aged 21–23 years) with normal or corrected normal vision faced a CRT with a background of 10 cd/m^2. One of the three chromatic stimuli (R, 570 nm; G, 535 nm; B, 445 nm; 5° × 5°) was presented for 2 s with an ITI of at least 10 s. Bipolar EEG was recorded from the occipital cortex (TC 0.3 s, Hi-cut at 60 Hz) together with the EOG (TC 5 s, Hi-cut at 60 Hz). Five out of ten subjects showed continuous alpha waves in EEG recordings even under an eye-opened state (Fig. 2a). These subjects showed a distinct alpha-blocking to chromatic stimuli without any

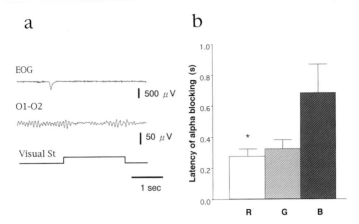

FIG. 2a,b. EEG alpha-blocking by chromatic stimuli. **a** In half of subjects (5/10), continuous alpha waves were observed even under an eye-opened state as revealed by eyeblink in the trace of electrooculogram (EOG), and distinct alpha-blocking was elicited by chromatic stimulus without any change in luminance. **b** Mean latency of alpha-blocking by blue was significantly longer than that by red ($t(37) = 2.274$, $*P < .05$). Mean with SE

changes in luminance and the latency was measured for each hue stimulation. As shown in Fig. 2b, mean latency of alpha-blocking was longer to blue as compared with those to red and green. This result is consistent with the earlier result that the RT was longer to blue stimulation. These results are consistent with a recent finding in monkeys that, in the brain, blue-on information is processed via a koniocellular pathway that has slower conduction velocities than do those devoted to the processing of red and green [5].

Effects of Hue on Discriminative Reaction Time

Discriminative RT was measured in the discrimination of one color stimulus from the others with a go-no go task, so that it includes the time required for discrimination of hue in addition to its detection time, which is mainly analyzed in the case of simple RT. Six undergraduate subjects were asked to press a button only to the target, that was previously told out of three chromatic stimuli (R, 635 nm; G, 548 nm; B, 463 nm). Discriminative RT was longer than simple RT in all of the chromatic stimuli. Discriminative RT was significantly longer to blue (Fig. 3a). Moreover, the difference in time, calculated by subtracting simple RT from the discriminative RT in each hue, was also longer to blue (R, 65.0 ± 15.3 ms; G, 86.1 ± 9.6 ms; B, 93.3 ± 14.6 ms). A nearly significant and strong tendency of difference was observed between the times to blue and those to red ($t(5) = 1.787$, $P = .134$). These results suggest that not only the detection but also the discrimination process for blue takes longer.

Right Hemispheric Dominance in Color Detection

It is generally known that the right hemisphere is dominant in visuospatial information processing [6]. In the present study, we examined whether or not hemispheric dominance exists in color detection. Human optic nerves originating from the nasal

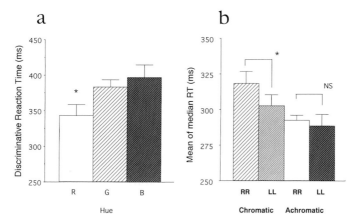

FIG. 3. **a** Discriminative RT for chromatic stimuli was significantly longer to blue than to red ($t(5) = 3.086$, * $P < .05$). Mean with SE. **b** RT by left hand to target in left visual field (*LL*) was significantly shorter than RT by right hand to target in right visual field (*RR*) only when the target was chromatic ($t(9) = 3.171$, * $P < .05$). Mean with SE

retina project to the contralateral visual cortex, while the nerves from the temporal retina project to the ipsilateral visual cortex. Pyramidal tracts from the motor cortex control contralateral hand movements; thus, the right motor cortex innervates the left hand and vice versa. Based upon these double crossed projections, we evaluated the time required for detection of color in each hemisphere by comparing RTs to visual stimuli presented to either the right or left visual field and responded to by the ipsilateral hand [7]. Ten undergraduate students with normal vision (age range 21–23 years, both genders, right-handed) were asked to attend two blocks of experiments. The target, either chromatic (R, 635 nm; G, 535 nm; B, 445 nm) or achromatic (12–20 cd/m^2), of a 2° circle was presented on a background of 10 cd/m^2. EOG was recorded to monitor the eye fixation. In the first experiment, the target was presented at the center of visual field and there was no significant difference between RTs by right and left hand. In the second experiment, the target appeared at 4° horizontally from the fixation point in either the right or left visual field. As shown in Fig. 3b, RT by the left hand to the chromatic target presented in the left visual field was significantly shorter than RT by the right hand to the target in the right visual field. On the other hand, there was no significant difference between the RTs to achromatic targets. These data show that the right hemisphere is dominant in the detection of color stimulus in human subjects.

Conclusions

A behavioral study using RT showed that blue information is uniquely processed in the brain. Simple RT was longer to blue as compared with red, yellow, or green. The results of EEG response were consistent with this finding.

Longer time for processing of blue information seems, at least partly, to be ascribed to longer processing time in the retina and subcortical pathway.

Results of discriminative RT show that discrimination time for blue is also longer. As the discrimination process is one of the functions of the color center, these findings suggest that processing of blue in the cortical visual center also takes a longer time. A comparison of reaction times between right and left hemispheres shows that the right hemisphere is dominant in the detection of color.

References

1. Donders FC (1969) On the speed of mental processes. Acta Psychol (Amst) 30:412–431
2. Schwartz SH (1995) Dependence of visual latency on wavelength: predictions of a neural counting model. J Opt Soc Am A 12:2089–2093
3. Finn JP, Lit A (1971) Effect of photometrically matched wavelength on simple reaction time at scotopic and photopic levels of illumination. Proc Annu Conv Am Psychol Assoc 6(Pt 1):5–6
4. Ueno T, Pokorny J, Smith VC (1985) Reaction times to chromatic stimuli. Vision Res 25:1623–1627
5. Martin PR, White AJ, Goodchild AK, et al. (1997) Evidence that blue-on cells are part of the third geniculocortical pathway in primates. Eur J Neurosci 7:1536–1541
6. Nebes RD (1973) Perception of spatial relationships by the right and left hemispheres in commissurotomized man. Neuropsychologia 11:285–289
7. Berlucchi G, Heron W, Hyman R, et al. (1971) Simple reaction times of ipsilateral and contralateral hand to lateralized visual stimuli. Brain 94:419–430

Inhibitory Mechanisms Underlying Stimulus-Selective Responses of Inferior Temporal Neurons

Hiroshi Tamura[1,3], Hidekazu Kaneko[2], Keisuke Kawasaki[1,3], and Ichiro Fujita[1,3]

Key words. GABA, Column, Area TE, Object recognition, Monkey

Introduction

Visual information regarding the shape and surface characteristics of objects is processed in the primate brain along the cortical pathway projecting to the inferior temporal cortex. Individual neurons in cytoarchitectonic area TE of the inferior temporal cortex respond preferentially to a range of complex visual stimuli such as shapes, shapes combined with color or texture, or complex images such as faces [1]. The prestriate areas V2 and V4, as well as the posterior part of the inferior temporal cortex, also contain a substantial population of neurons that respond better to shapes such as crosses and hyperbolic or polar gratings than to bars, edges, or linear gratings [2]. The stimuli necessary for strong activation of neurons in these areas are generally simpler than those that excite TE neurons. Object information carried by single neurons is integrated gradually into complex forms in successive areas. Both convergent/divergent projections between areas and local excitatory and inhibitory interaction within an area contribute to this process. Inhibitory neurons have been implicated in various sensory processing including the generation of orientation selectivity in the primary visual cortex [3]. Here we discuss the role of area TE inhibitory interneurons in the generation of complex stimulus selectivity.

Disinhibition Reveals Latent Inputs to TE Neurons

Removal of inhibition by local application of bicuculline, a γ-aminobutyric acid $(GABA)_A$ receptor antagonist, in the vicinity of a TE neuron modulates its stimulus-selective responses [4, 5]. Bicuculline application augments responses to some stimuli

[1] Graduate School of Frontier Biosciences, Osaka University, Toyonaka, Osaka 560-8531, Japan
[2] National Institute of Advanced Industrial Science and Technology, Tsukuba 305-8566, Japan
[3] Core Research for Evolutional Science and Technology, Japan Science and Technology Corporation, Toyonaka, Osaka 560-8531, Japan

and even unmasks responses to stimuli that do not elicit responses under normal conditions, drastically altering the stimulus selectivity of the neuron. These effects are observed for particular groups of stimuli, including those related to the originally effective stimuli and those that do not initially excite the neurons but activate nearby neurons. These findings indicate that TE neurons receive diverse stimulus inputs, and that some of these inputs are effectively masked by GABAergic inhibition to acquire their final response properties.

Inhibitory Neurons in Area TE Are Stimulus Selective

Stimulus-specific inhibition does not necessarily require inhibitory neurons with stimulus-specific responses. It can occur if inhibitory neurons with poor stimulus tuning shunt stimulus-specific excitatory inputs. Conversely, broadband inhibition can be mediated by stimulus-selective inhibitory neurons, if they are selective for different stimuli and converge onto the same target neuron. These possibilities may be separated by investigating the response properties of inhibitory neurons in area TE, and addressing directly whether responses of inhibitory neurons are stimulus selective and how their stimulus selectivity is related to that of their target neurons [6].

Inhibitory neurons can be identified by recording extracellular spikes simultaneously from multiple adjacent neurons and applying a cross-correlation analysis to the spike trains [7]. A neuronal pair (Fig. 1a) shows a trough shifted toward the positive direction from the 0-ms bin in their raw cross-correlogram (CCG), indicating an inhibitory influence of the reference cell (cell 1) on the other cell (cell 2). Disappearance of the trough in the shift-predicted CCG indicates that the trough in the raw CCG is due to neuronal interaction and not a stimulus-locked change in firing rates.

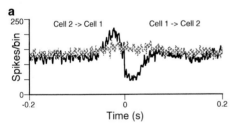

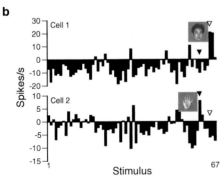

FIG. 1a,b. An example of a cross-correlogram (CCG) with a displaced trough between a pair of adjacent neurons in area TE. **a** CCG with a trough shifted from the 0-ms bin. The CCGs cover ±0.2 s with 0.4-ms bins. The raw (*solid line*) and shift-predicted (*dotted line*) CCGs are shown. **b** Response profiles of cell 1 and cell 2 to a set of 67 visual stimuli: 53 two-dimensional geometrical shapes (circles, squares, triangles, stars, gradation patterns, gratings, etc.) and 14 photographs of objects (banana, face, hand, syringe, etc.). *Open arrowhead* indicates stimulus 65, which evoked the strongest response in cell 1, and *closed arrowhead* indicates stimulus 61, which is the most evocative stimulus for cell 2

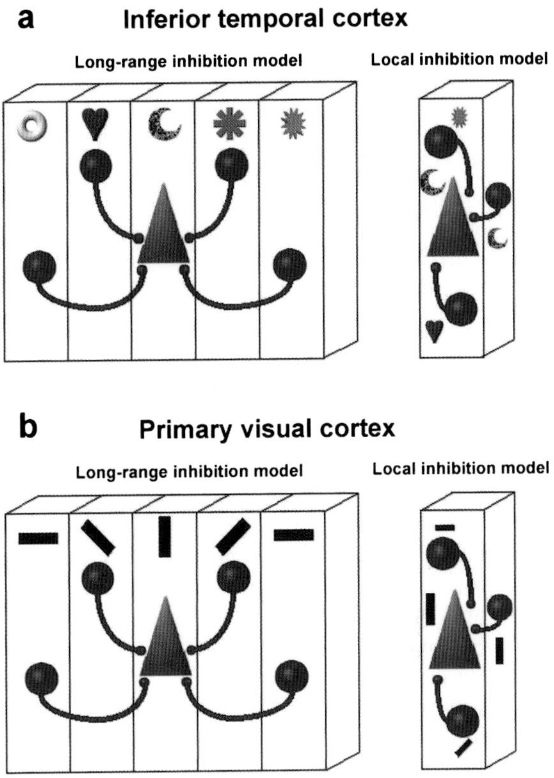

FIG. 2a,b. Long-range inhibition models and local inhibition
models for visual response
selectivity in area TE (**a**) and the
primary visual cortex (**b**). *Tri-
angle*, excitatory neuron; *circle*,
inhibitory neuron. In long-
range inhibition models (*left*),
inhibition originates from different columns, which encode
different visual features (**a**) or
orientations (**b**). In local inhibition models (*right*), inhibitory
neurons within the same
column provide inhibition. In
this model, neuronal selectivity
is not perfectly uniform within
a column

Putative inhibitory neurons thus identified show clear stimulus preferences. For example, cell 1 responds only to 4 of the 67 stimuli (Fig. 1b). For a population of cells tested, the number of stimuli that evokes responses greater than 25% of the maximal in inhibitory neurons (21% of stimuli in our stimulus set) is not different from those that elicit such response in excitatory neurons (23% of stimuli; $P > 0.05$, U-test). Inhibitory TE neurons are as stimulus selective as excitatory TE neurons and can provide stimulus-specific inhibition.

Inhibition Operates Between Neurons with Different Response Profiles

Neurons that exhibit a trough on their CCGs show overall profiles of stimulus prefer-ences different from each other. For example, an image of a human face (open arrow-head), to which cell 1 responds maximally, evokes inhibitory responses in cell 2 (Fig. 1b). Conversely, cell 2 responds maximally to an image of a hand (closed arrowhead), by which cell 1 is inhibited. The Pearson's correlation coefficient between the responses of the two neurons is -0.38 ($P = 0.002$).

We classify neuronal pairs into three groups, according to the shape of their raw CCGs: neuronal pairs with a trough shifted from the 0-ms bin with or without an accompanying peak (inhibitory linkage), neuronal pairs with a sharp peak displaced from the 0-ms bin without a trough (excitatory linkage), and neuronal pairs with a peak straddling the 0-ms bin without a trough (common inputs). Response correlation differs among them ($P < 0.001$, Kruskal Wallis test). The coefficients of response correlation for pairs with inhibitory linkage, excitatory linkage, and common inputs are -0.04 ± 0.22, 0.24 ± 0.25, and 0.18 ± 0.22 (mean $\pm$ SD), respectively. Response profiles of pairs with inhibitory interactions are less similar than those of pairs with common and/or excitatory inputs, suggesting that local inhibitory interactions operate mainly in pairs with different response profiles.

Long-Range Vs. Local Inhibition Models

Many TE neurons in a columnar local region have been shown to respond to the same or similar stimuli [1]. The present results, however, suggest that inhibitory neurons with different response profiles reside in the same column and suppress responses of their target neurons to a range of stimuli, thus shaping the stimulus selectivity of TE neurons (Fig. 2a, right).

Models for orientation selectivity of primary visual cortex neurons typically assume that all neurons in a columnar region respond to the same orientation, and that cross-orientation inhibition originates from neurons in columns with different orientation preferences (Fig. 2b, left) [8, 9]. Although most adjacent neurons in the primary visual cortex share a preferred orientation, there exist neurons with a wide variety of preferred orientations in a local region [10]. It remains an open question whether the latter neurons are inhibitory neurons and provide cross-orientation inhibition from within the same columnar region in a similar way as in area TE (Fig. 2b, right; see also [11]).

References

1. Fujita I (1993) Columns in the inferotemporal cortex: machinery for visual representation of objects. Biomed Res 14(S4):21–27
2. Kobatake E, Tanaka K (1994) Neuronal selectivities to complex object features in the ventral visual pathway of the macaque cerebral cortex. J Neurophysiol 71:856–867
3. Sillito AM (1984) Functional considerations of the operation of GABAergic inhibitory processes in the visual cortex. In: Jones EG, Peters A (eds) Cerebral cortex, vol. 2. Plenum press, New York, pp 91–117
4. Wang Y, Fujita I, Murayama Y (2000) Neuronal mechanisms of selectivity for object features revealed by blocking inhibition in inferotemporal cortex. Nat Neurosci 3:807–813
5. Wang Y, Fujita I, Tamura H, et al. (2002) Contribution of GABAergic inhibition to receptive field structures of monkey inferior temporal neurons. Cereb Cortex 12:62–74
6. Tamura H, Kaneko H, Kawasaki K, et al. (2002) Visual response properties of presumed inhibitory neurons in the inferior temporal cortex of macaque monkey. Abstr Soc Neurosci 32, Program No. 160.7
7. Perkel DH, Gerstein GL, Moore GP (1967) Neuronal spike trains and stochastic point processes: II. Simultaneous spike trains. Biophys J 7:419–440
8. Wörgötter F, Koch C (1991) A detailed model of the primary visual pathway in the cat: comparison of afferent excitatory and intracortical inhibitory connection schemes for orientation selectivity. J Neurosci 11:1959–1979

9. Somers DC, Nelson SB, Sur M (1995) An emergent model of orientation selectivity in cat visual cortical simple cells. J Neurosci 15:5448–5465
10. Maldonado PE, Gray CM (1996) Heterogeneity in local distributions of orientation-selective neurons in the cat primary visual cortex. Vis Neurosci 13:509–516
11. Matsubara JA, Cynader MS, Swindale NV (1987) Anatomical properties and physiological correlates of the intrinsic connections in cat area 18. J Neurosci 7:1428–1446

Key Word Index